WATER RIGHT

Conserving Our Water
Preserving Our Environment

TABLE OF CONTENTS

Preface

WATER Everywhere
Dr. H. Marc Cathey

Introduction

Chapter I / Page 3

The Scope of Water Problems: Quality, Quantity and Beyond

Chapter II / Page 7

Water Use and Conservation: Perception vs. Reality

Chapter III / Page 11

Environmental Benefits of Responsible Landscape Management

Chapter IV / Page 15

Economic Value and Benefits of Responsible Landscape Management

Chapter V / Page 18

Educational Needs and Opportunities for Water Conservation

Chapter VI / Page 21

Landscape Water-Conservation Techniques

Chapter VII / Page 27

Conservation-Aware Individuals Will Make the Difference

CASE STUDIES

Case Study 1: Never Underestimate the Importance of Plants to People / page 31

Case Study 2: 21st Century Landscape Water Use: A Global Perspective / page 33

Case Study 3: Soil-Water Issues Relevant to Landscape Water Conservation / page 36

Case Study 4: Refining the Concept of Xeriscape / page 39

Case Study 5: No Water Should Be "Waste Water" / page 42

Case Study 6: The Important Role of Science in Landscape-Ordinance Development / page 45

Case Study 7: Water Conservation on Golf Courses / page 48

Case Study 8: Homeowners Can Conserve Water with Low-Tech and High-Tech Solutions / page 52

Case Study 9: Maintaining Superior Landscapes on a Water Budget / page 55

Case Study 10: Communicating Water Conservation to a Community / page 57

APPENDICES

Appendix A: Indoor and Outdoor Residential Water Conservation Checklist / page 61

Appendix B: Landscape Water Conservation Ordinances / page 62

Introduction

Clean, abundant and affordable water does not exist in many parts of the world today.

There is mounting evidence that more people in more places ultimately will face severe water shortages, and available supplies in these venues will be highly contaminated or very expensive.

In light of these dire facts and forecasts, a publication addressing the use of water for landscaping purposes may seem extraordinarily short-sighted at best or highly selfish at worst.

But this publication is the result of very different concerns.

A vast array of plant, social and environmental scientists has documented that when landscapes are properly designed, installed and maintained, relatively small amounts of water are required to achieve substantial benefits. In return for the proper amount of water they require, sound landscapes provide functional, recreational and aesthetic benefits that advance immediate and long-term personal and social well-being. These landscapes also help to purify contaminated waters as they recharge available supplies.

We are just now beginning to understand that the price of eliminating landscapes in the name of water conservation can be high. For example, when properly maintained landscapes are absent, fires spread more rapidly, floods ravage larger areas and the accompanying erosion from both types of catastrophes further spoils the environment and water supply. Any real or perceived water savings gained by eliminating landscapes can prove fleeting indeed.

We have not seen the future, but we can expect that tomorrow will be different from today. Shortsighted, single-solution thinking will be replaced (even if slowly) with a more global, universal and synergistic approach to identifying problems and finding solutions.

Water-policy decision-makers, by their very position in society, can make a tremendously positive impact on lives, livelihoods and living conditions. This is particularly so if these officials expand their horizons courageously and innovatively by adopting forward-thinking new approaches to water use, conservation and quality.

This publication aims to encourage new thinking. Chapters I through VII summarize findings of scientists who have studied water quality and quantity problems and reached conclusions about the misconceptions surrounding them. These chapters spell out scientifically supported solutions that enlightened landscape water usage can offer. Case histories illustrate the benefits realized when water purveyors and users have cooperated to apply the best landscape water-conservation practices. The two appendices offer a practical water conservation checklist and a review of landscape conservation ordiances, followed by a set of principles for water conservation.

Chapter I

The Scope of Water Problems: Quality, Quantity and Beyond

Summary:

Water shortages and water-quality issues are global, not simply local. Emergence of these issues is a matter of "when," not "if." There is a need to both conserve and clean the world's water supplies. Solutions need to be based on site-specific determinants and have long-term considerations.

A child races along a long ribbon of pristine beach having played the day before chasing friends through sprinklers in her yard. Her diet consists of fine and refined foods as well as soft drinks, milk and the occasional imported, exotic fruit drink. With the turn of a tap she sees what seems to be an unlimited quantity of clean water rapidly flowing into the sink, shower or tub and then down the drain. She shows no trace of concern for where the water comes from or where it's going. The cost of getting the fresh water to her and purifying her wastewater never enters her mind.

Is there really a water-shortage problem?

In many parts of the developed world, parents may be less informed about water issues than their children because of relatively new efforts in many schools to increase water-conservation awareness.

At the same time, water is all but unavailable in an increasing number of developed and developing countries alike. Now is the time to question what has caused water

INCREASING POPULATION

LAND USE AND URBANIZATION

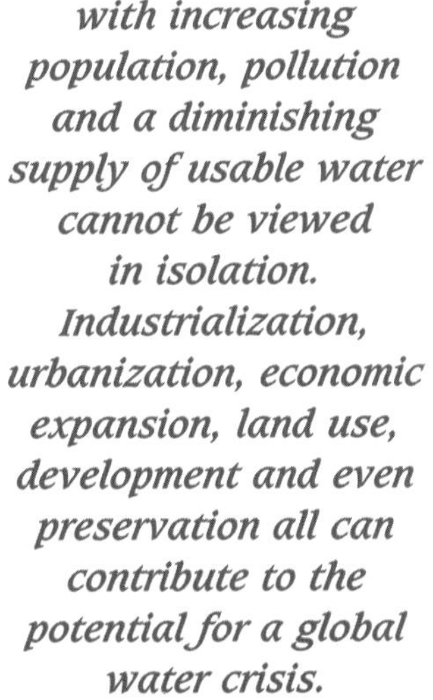
The problems and solutions associated with increasing population, pollution and a diminishing supply of usable water cannot be viewed in isolation. Industrialization, urbanization, economic expansion, land use, development and even preservation all can contribute to the potential for a global water crisis.

INDUSTRIALIZATION AND POLLUTION

ECONOMIC EXPANSION

shortages and why. We can also begin to ask what can be done to improve the situation for now and the future. The supply, source and use patterns of water are factors that can be easily identified and understood. Other aspects of water, however, are not recognized, understood or even considered. For example, it is universally agreed that water is in fact a limited but naturally recycling resource. We generally accept that 97 percent of the world's water lies in the oceans and seas and 2 percent is locked up as glacial ice, leaving only 1 percent available for human use. With only minor fluctuations, these percentages have remained unchanged for eons.

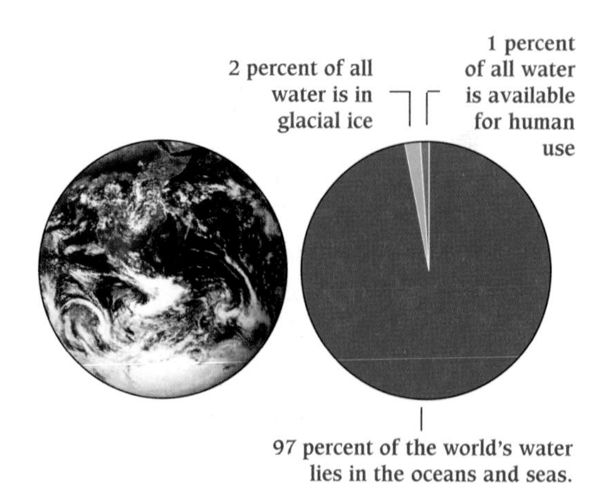

Conversely, human demand for water has risen at remarkable rates as a result of both increasing population and water use. Exacerbating the situation is the fact that the 1 percent of available water is being spoiled by various forms of pollution, thus reducing its use for our consumption.

In 1999, *The Futurist* magazine, in its top-10 forecasts, listed the belief of leading scientists, researchers and scholars that "water scarcity could threaten 1 billion people by 2025."

While not front-page news, the *Financial Times* of London reported an alarming array of little-understood water-related facts in January 1992:

- 80 percent of all diseases and 33 percent of deaths in developing countries are linked to inadequate water quantity and quality.
- Less than 2 percent of the 2 million tons of human excrement produced daily in cities around the world is treated, with the rest discharged into watercourses.
- The rate of pollution from industry and domestic consumers long ago surpassed the threshold of nature's recovery process, with alarming consequences for the natural environment and the health of city dwellers.

Human demand for water has risen at remarkable rates as a result of both increasing population and water use. Exacerbating the situation is the fact that the 1 percent of available water is being spoiled by various forms of pollution, thus reducing its use for our consumption.

- By the end of the 1990s, world water demand including requirements for waste dilution was expected to reach 18,700 cubic kilometers (4.9 quadrillion gallons) annually. This is almost half of the total global runoff. *Facing Water Scarcity* author Sandra Postel wrote as early as 1992 that 232 million people in 26 countries were living in water-scarce areas. She noted that Africa has the largest number of countries in which water is scarce, and she projected that by 2010 the number of Africans living in these countries would climb to 400 million, some 37 percent of the continent's projected population. More recently, Postel suggested, "The number of people living in water-stressed countries is projected to climb from 70 million to 3 billion by 2025."

Postel also reported that nine Middle Eastern countries face water-shortage conditions, and because the region shares many rivers, political tensions over water rights could worsen.

A future war over water is a very distinct possibility, according to Klaus Toepfer, director-general of the United Nations Environment Program. In a January 2000 interview published in the journal *Environment, Science and*

Technology he said, "Everybody knows that we have an increase in population, but we do not have a corresponding increase in drinking water, so the result in the regional dimension is conflict."

Reasons for Various Degrees of Water-Shortage Crises

Water-shortage crises don't happen only in low-rainfall or developing countries but in locations such as London, England; Melbourne, Australia; Seattle, Washington; and Reno, Nevada. All of these cities have faced various degrees of water shortages resulting in bans or restrictions on water use.

LONDON

MELBOURNE

All of these citiies have faced various degrees of water shortages relulting in bans or restrictions on water use

SEATTLE

RENO

Water shortages don't happen only in low-rainfall or developing countries but in locations such as London, Melbourne, Seattle, and Reno

These types of shortages arise for a wide range of reasons, not all of which are based on an actual shortage of water. In some cases development simply outpaces infrastructure. Homes and commercial structures are built, but local water service can't keep up because of supply, treatment capacity or quantity of pipe and pumping stations.

In other cases environmental concerns, regulations or legal decisions restrict the amount of water that can be used to serve an area's population.

Weather also creates water shortages when drought or insufficient snowfall accumulation fails to replenish reservoirs or when flooding contaminates supply.

Mechanical and structural shortcomings also cause water shortages. A main pump or pipe breaks, and water temporarily stops flowing into homes and businesses. In older systems more than 50 percent of treated water can be lost through major leaks.

Public water-supply system administrators widely agree that at least 10 percent to 15 percent of all treated water becomes "unaccounted for" in usage including fire-fighting. As necessary a water use as that is, it nevertheless causes a tremendous quantity of water to be lost between the treatment plant and the water meter.

Environmental water uses also are receiving greater attention and being given a higher priority for water allocation. Fish and wildlife preservation, as well as water-related sports activities such as boating and rafting, have in many instances required greater stream inflows than was previously considered necessary.

While this decreases the amount of water available for human consumption, it also helps to maintain the viability of natural areas, numerous species and recreational activities. As a result, another set of competing water needs arises.

Pollution is Another Very Significant Water-related Concern

For much of human history, people have chosen to live and work near easily accessible water to satisfy their daily consumption as well as agricultural and transportation needs. Today, development continues to be most prevalent near waterways, but with that comes a cost: water pollution. As noted earlier, more than 98 percent of the 2 million tons of human excrement produced daily in cities around the world is not treated but simply discharged into watercourses.

Industrial, commercial, agricultural and residential pollutants, as well as silt, also contaminate our groundwater and waterways directly and indirectly. Once water becomes contaminated, it may be impossible to purify or the cost too staggering to undertake — thus further

Fish and wildlife preservation, as well as water-related sports activities such as boating and rafting, have in many instances required greater stream inflows of clean water than was previously considered necessary.

reducing the available supply.

But the problems and solutions associated with increasing population, pollution and a diminishing supply of usable water cannot be viewed in isolation. Industrialization, urbanization, economic expansion, land use, development and even preservation all can contribute to the potential for a global water crisis.

No Simple or One-Size-Fits-All Solutions

Thus we need to be brave enough to ask serious questions about past and current water-policy solutions and even seemingly unrelated issues. It is commonly recognized that we can no longer afford to continue to pollute the air and water. But have we done enough to filter waters naturally before they run into streams or infiltrate groundwater sources?

We also need to think about how much of the globe we can cover with asphalt, cement and roofing materials, which impede the natural flow and recharging of our water supply and create energy-intensive heat islands. Further, what are the environmental and water-quality costs associated with reducing or eliminating water used on landscapes in the name of conservation? And we must consider how wise it is to engineer a system that rushes rainwater and snowmelt to the oceans rather than into or across areas where they can be useful.

Clearly there are no simple or one-size-fits-all solutions. But perhaps new thinking on the part of water-policy decision-makers, scientists, technicians and even the general public can reveal how we can use the world's 1 percent supply of fresh water more wisely so that future populations, regardless of numbers, can live comfortably. Perhaps we can even develop ways to increase the 1 percent figure without harming the environment.

We need to think about how much of the globe we can cover with asphalt, cement and roofing materials, which impede the natural flow and recharging of our water supply and create energy-intensive heat islands. Photo: some communities are helping to alleviate heat islands with landscaping.

Chapter II

Water Use and Conservation: Perception vs. Reality

Summary

Precise definitions of water types and water uses are essential. Some mandates have proven to increase, not decrease, water use. Conservation efforts can be most effective when consumers are well-informed.

When told we have to start conserving water, the average person might reasonably ask two apparently simple questions: "How much am I using now?" and "How much do I need to conserve?"

What Water Are We Conserving?

The perception is we all know what "water" we're talking about conserving — and that if we can all agree on how much is being "used," we can then determine how much to "conserve."

The reality is far different and much more complex, beginning with a definition of the simple word "use."

Unlike other renewable resources such as lumber and corn oil or non-renewable resources such as coal and oil, water is not used or consumed in the traditional definition of the words. More appropriately, it is stored in various forms and in various vessels. The forms can be solid, liquid or vapor. The vessels can be anything from the environment, such as glaciers, oceans, rivers and lakes, to pipes, tanks, cans and bottles and even plants, animals and humans. The reality is that a dinosaur may well have consumed the same water we drink today – because it has been recycled through the atmosphere time and time again. Just because that dinosaur drank the water, it was not irretrievably lost to today's use.

Scientists have concluded that the amount of water present on Earth has been relatively stable for eons, an estimated 290 million cubic miles of water.

Through a process called the "hydrologic cycle," precipitation in the form of rain, snow or hail generally equals the amount of water lost to evaporation. Because on a global average there is 30 percent more precipitation onto the land than evaporates from it, there is a potential annual net gain of approximately 9,000 cubic miles of water on the land every year. This is the fresh water that recharges our ground and surface water supplies, feeding the streams and rivers and eventually flowing into the oceans.

The Hydrologic Cycle

Although there is a potential annual net gain of approximately 9,000 cubic miles of water on the land every year, the paradoxical reality remains– water is not increased, it is only recycled.

The paradoxical reality is that while we are never going to exhaust our water supply, we cannot increase it – we can only recycle it.

What Type of Water Are We Talking About Conserving?

There is also a perception that we know what type of water we are talking about conserving.

But in reality, we lack agreement as to whether the water to be conserved should include all types: fresh water only, or salt water and effluent as well? Should all fresh water including ground and surface water be conserved, or only publicly treated and supplied water? Should conservation apply to all industrial, commercial, agricultural and domestic water use, or only to domestic outdoor use?

Confusion also can arise when it comes to distinguishing between off-stream and in-stream uses; between domestic, self-supplied and publicly supplied domestic and commercial water; and between direct, indirect or mixed-supply users. Furthermore, the term "personal use" can be understood either as what one individual actually consumes or requires for hygienic purposes, or it may incorporate the amount of the water used to provide that person with everything from drinking and bath water to the agricultural and industrial water used to produce an egg, car or newspaper!

Units of measure may be perceived as adding clarity, but in reality, they too can create confusion. Terms and abbreviations such as *million gallons per day* (Mgal/d); *acre-feet* (A-ft); *gallons per capita per day* (gpcd) and *100 cubic feet* (ccf) can be mind-numbing. Then consider converting everything to the metric system of cubic kilometers, liters, meters and hectares!

Who Owns the Water We Are Conserving?

Another perception/reality question relates to who "owns" water. Mark Twain once said of the western United States, "Whiskey is for drinking, and water is for fighting." He was right then, but the geographic application of his comment can now be considered to be global.

In some areas, you can "own" the water you can pump from beneath your property or whatever flows through it. But more and more that seems to be changing. Now greater consideration has to be given to "downstream" uses including those not only for human consumption or for production demands but also for environmental requirements. This results in the practice that requires water purveyors who withdraw water from a river to fully treat and return a certain percentage of that water to the river or face severe costs or penalties.

Many public water suppliers pay fees or have limits on the amount of water they can withdraw from a source (usually a stream or surface water) but are credited for the amount of treated water they return to that source. Under this arrangement, the public supplier has an obvious incentive to discourage outdoor water use because there is no way of accurately measuring how much water is being returned to the system, even if the costs for treating the returned water are extremely high. Thus there may be a systemwide disincentive to use effluent water for many of the same reasons.

Greater consideration has to be given to "downstream" uses including those not only for human consumption or for production demands but also for environmental requirements.

In such circumstances, the use of recycled or effluent water for industrial purposes or on landscapes must be discouraged, or there will not be sufficient return flow for downstream use. While the perception may be that recycled water usage can conserve potable water, the reality is that downstream needs may prohibit its consideration.

Who Directly Consumes the Highest Percentage of Water?

There is also a perception, in at least some circles, that homeowners directly consume the highest percentage of water and therefore they should be capable of conserving the largest amounts most easily. But this notion is debatable, depending again upon definitions. Cooling for thermoelectric generation and production agriculture requires the greatest amounts of fresh water, but domestic uses require the largest quantities of publicly supplied water.

Another reality is that water purveyors traditionally look first to their customers whose usage is highest when significant changes of volume in consumption are needed. Thus, it is not surprising that public water-system officials in expanding urban areas first look longingly at homeowners as the primary target for conserving publicly supplied water and then at the volume of water used for agriculture.

Ultimately, though, water-policy decision-makers usually conclude that by focusing their conservation attention on the greatest volume uses of water they will always achieve the largest savings. Thus, for publicly supplied water, domestic use is typically the first general target of conservation — and within that market outdoor water use has traditionally been the first segment of conservation-related activity, with considerable attention focused on turfgrass water use.

Conservation efforts typically unfold in predictable stages. First would come non-threatening, educational water-conservation messages in the media and as water-bill stuffers asking people to use less water.

As the need to conserve grew, so too would the severity of the plan, going from alternate-day outdoor watering, to turf-area limits, to outlawing some grass species in favor of others, and eventually to outright bans on the use of turfgrass. Alternative plants, defined either as "low-water using," or "native," would be prescribed or legislated for perceived conservation landscaping.

To one degree or another, some or all of this scenario has unfolded in locales including Marin County, California; Reno, Nevada; Atlanta, Georgia; Seattle, Washington; London, England; and parts of the Middle East.

New Thinking Is Starting To Emerge

While some of these measures may have had initial

Outdoor water use has traditionally been the first segment of conservation-related activity.

success, it is now being learned there is little scientific water-use data to support the listing of non-turf plants as "low-water using" or "native." In fact, while many such alternative plants may be able to survive on little applied water, they become high water users when people do irrigate them in an effort to develop a pleasing landscape.

It is also recognized that water-use rates can actually increase with alternate-day watering because people incorrectly believe they must water every other day without regard for the plant's actual need.

In addition to the fact that a variety of mandates intended to conserve water have not proven particularly effective, there is an increasing level of recognition that overall environmental quality can be dramatically diminished by such measures. Without trees and turfgrass to cool a surrounding area, "heat islands" can develop. These require increased use of air conditioners, which burn more and more energy that could be used in other ways or reserved for future use.

This causes pollutants that would otherwise be trapped in turf to be washed into waterways along with increased amounts of soil and silt, further defiling the downstream water supply or groundwater resources.

Research Findings About Urban Water Conservation

Residential End Uses of Water, an in-depth study conducted in 14 cities in the United States and Canada that was funded by the American Water Works Association Research Foundation and released in the year 2000, provides some intriguing findings about urban water conservation.

- The mix of indoor and outdoor water use is strongly influenced by annual weather patterns. As expected, sites in hot climates like the Phoenix area (including Tempe and Scottsdale) had a higher percentage of outdoor use (59-67 percent), while sites in cooler, wetter climates like Seattle, Tampa and Waterloo, Ontario, had much lower percentages of outdoor use (22-38 percent).
- 10 percent of homes were responsible for 58 percent of the leaks found. Households with swimming pools have 55 percent greater overall leakage on average than other households.
- Leakage is found to be generally lower for households that use drip irrigation or use a hand-held hose for watering as well as for those who have reported taking behavioral and technological actions to conserve water outdoors.
- Because outdoor water use is more discretionary than indoor uses, outdoor use can decline more rapidly when prices rise.
- Homes with in-ground sprinkler systems use 35 percent more water outdoors than those without in-ground systems.
- Households that use automatic timers to control their irrigation systems used 47 percent more water outdoors than those without timers.
- Homes with drip-irrigation systems use 15 percent more water outdoors than those without drip irrigation systems.
- Households that water with hand-held hoses use 33 percent less water outdoors than other households. Households that maintain gardens use 30 percent more water outdoors than those without a garden.

Perhaps most remarkable was this finding: The low water-use landscape group (xeriscapes) actually received slightly more water outdoors annually than the standard landscape group because of homeowners' tendency to overwater. A similar result also was documented in a 1998 Arizona State University study funded by the U.S. Environmental Protection Agency.

Yesterday's perceptions are being challenged with new information, and as a result, the potential exists for new realities. Chief among them is that water-policy decision-makers will realize the importance of clearly defined and understood terms, conditions and data.

Without clarity, there will be confusion. And confusion often leads to chaos, not conservation.

What is 1 inch of water?

One inch of water a week is generally recommended for maintaining a viable landscape including vegetables, turf, trees and flowers. But what is 1 inch of water?

The following conversions help make this clear.

1 inch of water (applied or rainfall)

- on 1,000 square feet = 624 gallons or 5,200 pounds
- on 1 acre = 27,200 gallons or 200,000 pounds
- on 1 square mile = 17.4 million gallons or 145 million pounds

1 gallon equals

- 128 fluid ounces, 8.337 pounds, 3.782 kilograms
- 15,100 drops, 16 cups, 8 pints, 4 quarts
- 231 cubic inches, 0.2337 cubic feet
- 0.83262 British or Imperial gallon
- 3,785.4 milliliters or cubic centimeters

1 cubic foot equals 7.48 gallons, 62.4 pounds

1 cubic yard equals 202 gallons, 1,685 pounds, 764.5 liters

1 cubic meter equals 264.2 gallons, 2,002 pounds

1 acre-foot (12-inch depth across 43,560 square feet) equals 325,851 gallons, 2.7 million pounds

Chapter III

Environmental Benefits of Responsible Landscape Management

Summary:

Long-term health issues need to be incorporated into water-conservation plans. Emphasis placed on landscape water conversation should be balanced by environmental benefits. People are the chief cause of wasted water and thus, the greatest hope for effective conservation.

In the name of public health and safety, or even as a political statement, landscapes in some areas have been altered to the point where the environmental damage is tremendous. Perhaps the most devastating example is now coming into public view.

Political and Environmental Motivations for Restrictive or Alternative Landscape Recommendations

During the Cultural Revolution in the People's Republic of China during the 1960s, Chairman Mao and his followers deliberately removed all vestiges of what was considered to represent "Western civilization." This included the virtual elimination of all turfgrass areas and many types of trees.

More than three decades later, the human and environmental price of this action is just being thoroughly recognized and calculated. The lack of shade trees and turf causes cities to bear the burden of "heat islands" – which are 10 degrees to 30 degrees hotter than outlying rural areas. Further, when turfgrass is removed, the amount of smog and dust in the air increases because there are not sufficient numbers of plants to hold down the dust and trap particulate pollutants. Dust carries disease, bacteria and viruses, incidences of which rise. Moreover, the lack of turf also increases erosion, which raises levels of pollution and damages water quality in ponds, streams, rivers and lakes.

Throughout China today there is a tremendous effort underway to repair the landscape and with it the environment and public health. Tree and turf areas are being expanded as quickly as possible before further damage can be done, but it will take several decades and many generations before the effort will be completed and start to yield results.

While perhaps most dramatic and widespread, China is not alone in making water-related or landscape decisions that have proved less then wise.

During the Chinese Cultural Revolution the government virtually eliminated all turfgrass areas and many types of trees. Today there is a tremendous effort underway to repair the landscape. Dr. Paul E. Rieke, retired Professor of Turfgrass Management in Crop and Soil Sciences at Michigan State University, served as consultant to the Chinese government.

In Western Australia, large numbers of trees were planted on sandy soils surrounding reservoirs in the hope of reducing erosion caused by wind and water. Recently it was determined that the trees were actually consuming huge amounts of reservoir water, so they are being removed and replaced with turfgrass.

Many communities have added areas of turfgrass, plants and trees to reduce high levels of heat and glare during the day.

Throughout the desert southwest of the United States, during a terrible extended drought in the late 1980s and early 1990s, landowners were told to remove turfgrass and replace it with a variety of other plant materials. Wildfires swept through many areas, consuming vegetation and many homes. But those people who had maintained significant amounts of turf around their homes suffered very little fire damage.

The inner-city heat islands noted in China are common to urban areas that lack the cooling effect of evapotranspiration and shade that are provided by properly watered growing trees and turf. As a result, urban areas tend to suffer much higher temperatures than nearby rural areas. Not only do the buildings, streets, sidewalks and paved plazas reflect tremendous levels of heat and glare during the day, but they also retain significant amounts of heat energy during the night, so cooling seldom occurs in built-up areas. When rains do come, the water is rushed into a highly engineered sewer system where it must be treated and released on its journey to the oceans, rather than being allowed to naturally soak into the soil, replenishing soil moisture, recharging the groundwater supplies or flowing naturally into streams, filtered by the roots of trees and turf.

Lanscape Codes, Ordinances Influenced by Profit Motive

Beyond political or environmental motivations for restrictive or alternative landscape codes, ordinances and recommendations lies the profit motive.

The myopic or single-focus approach to domestic water conservation typically calls for strict limits on lot sizes and landscape components. These limits tend to restrict amounts of turfgrass, favoring instead plants incorrectly identified as "low water-using" or "native."

Such an approach allows builders and developers to expand their businesses, along with suppliers of building materials and home furnishings. Some banks and lending institutions also favor this approach to increase the number of mortgages; if lots are smaller due to landscape restrictions, more homes will be built. And because alternative or restrictive landscapes may require or are best designed by professional landscape architects and installed by professional landscape contractors, some firms within this industry quietly support this approach, all in the name of water conservation.

The fact remains, however, that residences use only 8 percent of the total freshwater withdrawals and up to 60 percent of the domestic supplied water. But new residential development requires the water-supply system to satisfy the expanded commercial and industrial needs of the increased population as well. Thus building more homes on smaller lots doesn't increase water conservation because more water is required for other related, non-residential uses.

The planned water-savings projected for reduced landscape water use will be offset by the new water demand for non-landscape purposes. The result can be an increasingly dense population living in an area that is hot, dry and dusty due to inadequate evapotranspiration. Ultimately, a new water-resource problem will have been created.

The Benefits of a Well-maintained Green Landscape

Because the benefits of well-maintained green landscapes (including turfgrass lawns) are not widely understood, this area has become an easy target for water conservation. This is especially so in light of the highly visible use of water on lawns and the all-too-frequent examples of waste: water flowing from lawns onto streets and other hard surfaces; irrigation systems running during downpours; and broken sprinkler heads spewing water like an open fountain. Absent any recognized social or environmental benefit of turfgrass, restricting lawn watering or lawns themselves have been easy answers — but perhaps not wise ones.

Published in the *Journal of Environmental Quality,* the research of Dr. J.B. Beard and Dr. R.L. Green provides strong evidence of the many important benefits provided by turfgrass[1] and divides them into three categories as follows.

Diagrammatic summary of benefits derived from turfgrass

Benefits of Turfgrass

Functional
- Soil erosion control
- Dust prevention
- Rain water entrapment & ground water recharge
- Solar heat dissipation
- Glare reduction
- Organic chemical/ pollutant entrapment and degradation
- Air pollution control
- Nuisance animal/ pest reduction
- Fire prevention
- Security—visibility
- Environmental protection

Recreational
- Low cost surfaces
- Physical health
- Mental health
- Safety cushion
- Spectator entertainment

Aesthetic
- Beauty
- Quality of life
- Mental health
- Social harmony
- Community pride
- Increased property values
- Complements trees and shrubs in the landscape

Erosion control

Sports safety

Community pride

Functional benefits: soil erosion and dust stabilization; groundwater recharge and surface-water quality; organic chemical decomposition; soil improvement and restoration; heat dissipation and temperature moderation; and noise abatement and glare reduction. Other benefits also include decreases in noxious pests, allergy-related pollens and human exposure to disease; safety in vehicle operation and equipment longevity; security for vital installations, reduced fire hazards and improved wildlife habitat.

Aesthetic benefits: improved mental health via a positive therapeutic impact; increased property values; a sense of community pride and social harmony; quality of life; general beauty and a complement to trees and shrubs in the landscape.

Recreational benefits: low-cost surfaces; physical health; mental health; safety cushion and spectator entertainment.

In addressing the issue of water conservation, Beard and Green wrote:

• Trees and shrubs can use more water than turfgrass

"If one compares the evapotranspiration studies that are available, typically trees and shrubs are found to be higher water users than turfgrasses on a per-unit land-area basis."

• Drought-resistant plants are not necessarily low water users

"Much confusion has arisen from the low water-use landscape plant lists from (some) xeriscape groups that have been widely distributed. The lists are based on the

[1] J.B. Beard is a former member of the Department of Soil and Crop Sciences, Texas A&M University, and currently heads the International Sports Turf Institute. R.L. Green is a member of the Department of Botany and Plant Sciences, University of California-Riverside. Beard and Green published *The Role of Turfgrasses in Environmental Protection and Their Benefits to Humans* in 1994. The study was then published in the May-June 1994 issue of the *Journal of Environmental Quality*. JEQ is published by the American Society of Agronomy, the Crop Science Society of America and the Soil Science Society of America.

incorrect assumption that those plants capable of surviving in arid regions are low water users, when these plants typically are only drought-resistant.

"When these species are placed in an urban landscape with drip or other forms of irrigation, many can become high water users. This occurs because the physiological mechanisms controlling evapotranspiration and drought resistance are distinctly different and cannot be directly correlated within a plant species or cultivar."

• Adjacent trees and shrubs reap benefits of turfgrass watering

"When turfed areas are irrigated, the adjacent trees and shrubs also are being irrigated as a result of the multitude of shallow tree and shrub roots that concentrate under the irrigated turf area. Thus, when a homeowner is irrigating the lawn, most of the adjacent trees and shrubs also are being irrigated.

• Turfgrass' brown color during drought periods is entirely normal

"Numerous turfgrass species are capable of ceasing growth, entering dormancy and turning brown during summer drought stress, but they readily recover once rainfall occurs. Some people incorrectly assume turfgrasses must be kept green throughout the summer period to survive, and thus will irrigate. Many trees drop their leaves during summer drought stress or during the winter period when only brown bark remains.

"What then is wrong with a tan to golden-brown turf during summer droughts if one chooses not to irrigate? If water conservation is a goal, then a dormant turf uses little water, whereas certain trees and shrubs may continue to remove water from lower soil depths."

• Water conservation can be achieved with low water-use turfgrasses

"In summary, there is no valid scientific basis for water conservation strategies or legislation requiring extensive use of trees and shrubs in lieu of turfgrasses. Rather, the proper strategy based on good science is the use of appropriate low water-use turfgrasses, trees and shrubs

Plant breeders strive to develop cultivars that require less moisture. In this photo, the center plot is an experimental bluegrass with exceptional resistance to summer drought and heat stress, compared to surrounding plots. Genetic improvement in turfgrasses offer the potential for significant water savings.

for moderate-to-low irrigated landscapes, and similarly to select appropriate dehydration-avoidant turfgrasses, trees and shrubs for non-irrigated landscape areas.

"The main cause for excessive landscape water use in most situations is the human factor. The waste of water results from improper irrigation practices and poor landscape designs, rather than any one major group of landscape plant materials."

Beard and Green conclude: "It is critical to educate the general public that the darkest green turf, which many people strive for, is in fact not the healthiest turf. A medium-green turf with moderate growth rate will have the deepest root system with less thatching, reduced disease and insect problems, and increased tolerance to environmental stresses such as heat, drought, cold and wear."

Thus, proper establishment and maintenance of landscaping, including turfgrass, clearly yield significant environmental and social benefits. However, there is also a clear need to move beyond the notion of "beautiful landscapes" to public education. Consumers must be helped to understand, appreciate and put into use scientifically based principles that result in environmentally positive landscapes that incorporate water conservation to the greatest degree possible.

Chapter IV

Economic Value and Benefits of Responsible Landscape Management

Summary:
Landscapes have considerable direct and indirect economic values. People spend large sums of money on landscaping to enhance the personal and economic value of their homes. The presence or absence of well-maintained landscapes significantly affects local economies.

Like a work of art, the value of a landscape is in the eye of the beholder.

We've all heard about a priceless old-world masterpiece being discovered in an attic because its owner perceived it to be worthless. But at auction the rejected piece of artwork fetches millions of dollars, verifying that it does indeed have value.

In the United States alone, homeowners spent $50.9 billion to install, improve and maintain their landscapes and gardens in 1999 – an increase of 8.3 percent over 1998 when they spent just under $47 billion

In a somewhat similar way, some people conclude that landscapes have no perceptible value and it is therefore appropriate to restrict or ban landscape-related outdoor water use.

Landscapes Have Considerable Direct and Indirect Economic Value to Many Segments of the Community

But landscapes do indeed have considerable direct and indirect economic value, not only to property owners and suppliers of landscape-related goods and services but also to the community at large.

In 1999 the Gallup Organization conducted a survey for the National Gardening Association to determine the direct economic value of landscapes. The survey found that in the United States alone, homeowners spent $50.9 billion to install, improve and maintain their landscapes and gardens in 1999 – an increase of 8.3 percent over 1998, when they spent just under $47 billion. NGA figures further reported that Americans spent more than $213 billion between 1995 and 1999 on their yards and gardens. The magnitude of spending indicates the value people place on landscaping.
In the same Gallup survey, a representative sampling of U.S. homeowners was asked to identify the most important benefits of a residential or commercial property having a well-maintained lawn and landscape. They ranked the benefits as follows.

- Beauty and relaxation for the family, employees or visitors (54.0 percent)
- Reflects positively on its owner (53.2 percent)
- Comfortable place to entertain, work at or visit (47.4 percent)
- Increased real estate market value (44.1 percent)
- Helps to beautify the neighborhood (43.3 percent)
- Provides a safe, high-quality play area for children (36.7 percent)
- Provides an exercise area for pets (21.3 percent)

Landscaping offers the best return on investment when making home improvements

- Helps to purify the air (19.9 percent)
- Helps to cool the air (17.8 percent)
- Provides a natural water filter to protect water quality and the environment (13.5 percent)

While aesthetic reasons topped the list of perceived benefits, it is significant that increased real estate market values ranked higher than either child safety or several environmental benefits.

It is important to recognize that placing high values on landscaping is not a recent phenomenon. Perhaps one of the most definitive reports on the topic, *The Value of Landscaping* was published in 1986 by the Nursery Products Division of Weyerhaeuser, a major international forest products company.

The report continues to provide an excellent benchmark, with the following among its findings:

- A Gallup Organization poll indicated that new home buyers and buyers of previously owned homes believe that landscaping adds nearly 15 percent, on average, to a home's value or selling price.
- Real estate appraisers, however, rate the value of residential-property landscaping at 7.28 percent and commercial-property landscaping at 6 percent.
- The April 1986 issue of *Money* magazine reported: "Landscaping improvement has a recovery value of 100 percent to 200 percent if it is well done and harmonizes with foliage nearby. This compares to a recovery value of a kitchen overhaul of 75 percent to 125 percent; a bathroom [renovation], 80 percent to 120 percent; a new deck or patio, 40 percent to 70 percent; and a swimming pool, 20 percent to 50 percent."
- There is a direct correlation between the price paid for a home and the influence of landscaping in the buying decision. Buyers of higher-priced homes are more influenced by landscaping.

Growth of Floriculture and Horticulture

As a result of the public's interest in improved landscapes (achieved either on a do-it-yourself or hire-it-done basis), a segment of the economy has expanded to meet these demands.

The U.S. Department of Agriculture's Economic Research Service reported in October 1999 that "Floriculture and environmental horticulture is the fastest-growing segment in U.S. agriculture in grower cash receipts [for 1998], averaging 9 percent annual growth." Horticulture ranks sixth among commodity groups in U.S. agriculture in terms of grower cash receipts behind cattle and calves, dairy products, corn, hogs and soybeans.

A year 2000 study released by the California Green Industry Council reports: "The green industry in California is an economic powerhouse. It's one of California's largest industries. University studies have consistently found this dynamic, growing industry represents over $12 billion in sales and 130,000 employees. What's more, landscapes cover over 1.6 million acres in California – making our back yards one of California's largest and most valuable resources."

A 1994 University of Florida study that examined turfgrass in the state reported: "There was about 4.4 million acres [of turfgrass], with 75 percent of this area in the residential household sector. Turfgrass-industry employment was 185,000 full-time and part-time workers, or 130,000 full-time equivalents. Value added to Florida's economy by all sectors of the turfgrass industry totaled $7.3 billion."

While temporary outdoor watering restrictions will not cause a severe loss of established plant material or landscape integrity and value, extended restrictions or long-term bans can be devastating not only to the plants but also to a large segment of the area's economy and possibly even the environment.

The consequences of a long-term outdoor watering ban can extend well beyond the loss of beautiful landscapes and the environmental benefits that they provide. Many businesses (including those not associated directly with the landscape industry) can lose income and profits, causing employee cutbacks and layoffs. In areas where well-maintained landscapes attract a wide variety of short- and long-term visitors, tourism revenues can drop. Commercial and residential development can also decline as fewer people want to move into an area that cannot provide its citizens with what they view as a reasonable amount of water.

In areas where well-maintained landscapes attract a wide variety of short- and long-term visitors, tourism revenues can drop when droughts and water bans affect water use.

More and more people are gaining an appreciation for the economic and environmental values of properly designed and maintained landscapes. As a result, they are acting to ensure that sufficient water will be available for landscape maintenance, particularly when superior water-use education programs are consistently encouraged and available.

Who feels the impact when landscape watering restrictions or bans are put into place?

Depending on the degree of the restrictions, everyone from a high school student working part time selling lawn mowers to the owner of a landscape service or irrigation company can be affected. The impact can be economic, aesthetic and even environmental.

DIRECT IMPACT

- Homeowners (single-family dwellings)
- Apartment renters and condo owners (multi-family dwellings)
- Public and private airports, churches, cemeteries
- Golf courses, parks and playgrounds, sports fields
- Tourism
- Commercial operations (owners and employees)
- Feed/seed stores
- Gas stations
- Hardware stores
- Greenhouses, nurseries, and garden centers
- Home centers
- Irrigation systems (manufacturing and installation)
- Mail-order firms
- Mass merchandisers
- Production nurseries
- Supermarkets and drug stores
- Turfgrass sod farms
- Landscape professionals (owners and employees)
- Architects and designers
- Contractors
- Golf course superintendents
- Groundskeepers
- Lawn-care operators
- Sports field managers

INDIRECT IMPACT
(as a result of related lost sales, unemployment, etc.)

- Material sales and delivery
- Service providers (cafes, dry cleaners, service stations, etc.)
- Sales and use taxes
- Trucking and other transportation

Chapter V

Educational Needs and Opportunities for Water Conservation

Summary:

Science-based education is essential to successful water-conservation programs. Climatic differences will necessitate localized water-use recommendations. Outdoor water use includes pools, fountains and water features in addition to lawns. Inappropriate watering practices waste more water than any single plant group. Immediate, constant and consistent eduction programs will prove most effective.

If you are planning for a year, sow rice...
If you are planning for a decade, plant trees...
If you are planning for a lifetime, educate a person.

Chinese proverb

If you always do what you always did,
You'll always get what you always got.

Modern proverb

"At the desk where I sit, I have learned one great truth. The answer for all our national problems— the answer for all the problems of the world—comes to a single word. That word is 'education.' "

Lyndon B. Johnson

The world population is increasing, and with it the demand for water use. Pollution, even with increased awareness of its costs, continues to spoil parts of our water supply. Thus there can be no question that changes, perhaps very dramatic changes, will have to be made in the way water is used. Education about proper water use and conservation is critically important, particularly as it relates to outdoor water use.

Outdoor Water-use Conservation Programs

In the United States, indoor water use per person is relatively constant across all geographic and social lines, and evidence shows it may even be declining slightly, according to the recent American Water Works Association Research Foundation report, *Residential End Uses of Water.* However, outdoor water use varies according to climate. AWWA estimates that hot climates have a higher percentage of outdoor water use, ranging from 59 to 67 percent, than cooler climates, with 22 to 38 percent.

Because outdoor water use clearly represents the greatest opportunity for residential water savings, science-based education is key to conservation efforts – which can be undertaken by green-industry organizations, water purveyors and municipalities, government agricultural and horticultural advisers, educators, researchers and schools. If conservation programs are not based on the best available knowledge and technology, they likely will not only fail to achieve the desired water savings but also will discourage homeowners from undertaking future efforts if the initial ones fall short.

When contemplating outdoor water-use conservation-education programs, it will be important to localize recommendations according to specific climatic forces.

World Population Growth

1900—1.6 Billion

1950—2.5 Billion

2000—6.1 Billion

Population and water use are increasing, and yet the fact remains, only 1 percent of the world's water source is available for human use.

Homeowners in hot, dry climates use more water outdoors than those living in cool, wet climates for a variety of reasons. It's important to recognize that the term "outdoor water use" includes pools, spas and water features that are much more popular and tend to be larger in hotter, drier climates.

The term "outdoor water use" includes pools, spas and water features that are much more popular and tend to be larger in hotter, drier climates.

Evaporative losses from uncovered pools and other water features are as great as or greater than they are from planted landscape areas. In regions where pools are maintained year-round and plants go dormant (either because of heat, life cycle or cold), the evaporative losses from these features can be greater still.

Plants in hot, dry areas require greater amounts of applied water because the transpiration that takes place is many times greater than in cool, wet areas, and the plants have no other way to restore that lost moisture.

Two studies have indicated that water-conservation efforts based on xeriscape-style landscaping have not delivered on their promise. The AWWA study reported, "A comparison of average annual outdoor consumption resulted in the finding that the low-water-use landscape group actually used slightly more water outdoors annually than the standard landscape group."

Further, an Arizona State University study found that "xeriscapes in Phoenix and Tempe on average received at least 10 percent more water than traditional landscapes consisting of turf and other so-called 'high water-use' plants."

Plants Don't Waste Water, People Do

An apt phrase for any landscape water-conservation effort would seem to be: "Plants don't waste water, people do"— this is supported by the conclusions of researchers J.B. Beard and R.L. Green. In *"The Role of Turfgrasses in Environmental Protection and Their Benefits to Humans,"* they wrote: "The main cause for excessive landscape water use in most situations is the human factor. The waste of water results from improper irrigation practices and poor landscape designs, rather than any one major group of landscape plant materials."

Major Water Topics for Public Education Programs

The following major topics should be incorporated in all public programs on landscape water conservation:

Landscape design that incorporates the intended purposes to be served by the area as well as the climatic conditions and the desired level of maintenance. This would also incorporate as many "water-harvesting" features as possible and reduce or eliminate severe slopes or hard surface materials that create "heat-islands" and increase the load on air conditioners.

Plant selection that is based on actual (not anecdotal) water use, climatic conditions, and end-user requirements and individual desires.

Soil preparation that is based on soil-test results for all planting areas is essential to any water-conservation program. Beginning with properly prepared and appropriately amended soil will maximize water penetration.

Landscape maintenance that brings to the user continually updated information, techniques and tools. Major categories in this area would be:

Irrigation: Regardless of system used, know the minimal water requirements for all plants and planting areas and how this may change throughout the four seasons. Perform routine maintenance on the irrigation system to optimize efficient and uniform operation. Incorporate the latest technologies available such as ET information, controllers, rain/wind shutoff devices and changing automatic controllers to match the plant's seasonal water needs. Select watering times that maximize water availability to the plant and minimize evaporation or drift losses.

Turf should not be irrigated on narrow strips of land or areas that are difficult to water.

Fertility and pest management: A vigorously growing plant is its own best protection. This can best be achieved by proper watering and fertility, relying on pesticides only as needed and only after properly identifying the weed, disease or insect. Excessive fertilizing can harm most plants by creating excessive new growth and lushness, while it also increases the need to clip and prune and then dispose of this unnecessary yard waste. Over-fertilized plants typically also use water less efficiently. Finally, care should be taken with all fertilizers and pesticides to ensure that proper amounts are used to avoid runoff or leaching, which may pollute supplies of ground water or surface water.

Mowing, trimming and pruning: Regardless of the plant type, these important practices should be adjusted according to the season and the plant's specific needs.

Landscape quality standards: These should accept natural seasonal variations. Just as leafless deciduous trees from autumn to spring are natural, lighter than dark green turf in summer is also natural and actually results in stronger, healthier turf.

Outdoor Water Conservation Education Sources

Innovation in outdoor water conservation education does not have to be expensive and can tap a number of local resources. Landscape architects and contractors (either individuals or their professional associations) are excellent advocates of proper landscape techniques, as are wholesale and retail plant-material businesses including producers of turfgrass sod and nursery stock. Within the United States, extension specialists are also an exceptionally valuable resource for information and education. Water retailers and wholesalers can easily be effective and efficient sources of high-quality water-conservation information.

Multiple means to disseminate this information are available. Among the more highly successful approaches have been landscape workshops conducted by parks and recreation departments as well as builders and developers. School programs, offering information at all grade levels (with take-home literature) have proven successful, as has extensive involvement of the area's local newspapers, television and radio stations.

While the content of these programs will need to be customized for local conditions, the form and format do not have to be. Excellent educational success has been achieved with approaches including official proclamations from mayors and governors, utility-bill stuffers and public-service announcements geared to schools and garden clubs. Local garden and nursery centers as well as botanical gardens and arboreta can establish demonstration sites that show as well as tell people how to incorporate water conservation into their landscapes and other outdoor features.

The three most essential elements to a successful water-conservation education program are that it be initiated sooner rather than later; that it be constant; and, that it be consistent.

Chapter VI

Landscape Water-Conservation Techniques

Summary:

When given information and technology, people will make the effort to conserve. Plant selection will have less impact on water use than either irrigation or soil preparation. Water budgets give consumers more options and personal choice, while providing an effective way to conserve water.

Outdoor water-conservation measures typically focus on reducing or eliminating landscape water use. But implementing new scientific findings and advanced technology and general education can go a long way toward achieving the same end, just as these methods have proved successful in conserving water indoors. People will act to conserve water and improve the environment when properly informed of and motivated by the best scientific knowledge and technology.

Rather than attempting to regulate or ban a specific water use, the water budget technique leaves the determination in the hands of the rate-paying water user.

An Individual's Right to Choice

Numerous water-use studies have documented that depending on an area's climate, residential outdoor water use can account for between 22 percent to 67 percent of total annual water use. Clearly, this represents a vast opportunity for conservation. But in order to maintain an individual's right to personal choice and to maximize the positive environmental benefits of landscaping, a variety of factors need to be addressed taking location into account. When dealing with living plants, a one-size-fits-all solution will not be effective. But proven advanced scientific landscape water-conservation principles and practices do exist, and these can be modified and refined according to area-specific needs.

The Need for Clear and Careful Definitions

First, however, potential targets for widespread outdoor water conservation should be clearly and carefully defined. Too often, a narrow definition focuses exclusively on landscape water use. Narrow definitions often overlook potentially high water-use elements such as swimming pools or other water features, whose evaporative losses are as great as or greater than those from landscape application.

Other non-landscape outdoor water uses include washing cars, driveways, sidewalks and siding — and even some types of children's water toys. In addition to conserving in these areas, many techniques can be applied in the area of landscape water conservation as well.

The Water Budget Program

One of the most effective techniques is based on the practice of advising people how much water they can use, rather than telling them how they must use it. Termed a "water budget" or "water allocation" method, water providers establish a series of escalating allocation/pricing tiers so that every unit of water (i.e., 1,000 gallons), in excess of a base quantity, costs more than the previous unit.

Because outdoor water use can be measured and priced higher, people adjust their end uses according to their personal desires and financial concerns. This approach eliminates the need for contentious public hearings on landscape ordinances and the development of debatable plant lists, as well as the potential for draconian enforcement practices and so-called "water police."

Water budgets, for both indoor and outdoor water use, encourage individual freedom of choice and allow artistic expression on the part of homeowners and landscape designers. Rather than attempting to regulate or ban a specific water use, this technique leaves the determination in the hands of the rate-paying water user. As we've seen when gasoline prices rise, individuals can quickly adjust their use patterns. The same holds true for water.

Once purveyors decide on how to allocate and price water, they then have the very important role of assisting in the development and distribution of scientifically based educational materials on water conservation. Again, successful indoor conservation practices can be easily converted to outdoor water conservation, particularly as it relates to landscape water use.

The Two-Track Strategy

Moving from general to more specific landscape water-conservation recommendations, a two-track strategy that emphasizes different approaches for existing landscapes and newly planned and installed landscapes may be most effective, as follows. (For a comprehensive listing of indoor and outdoor water-conservation techniques, see Appendix A.)

I. Existing Landscape Areas

A. Pre-drought/pre-maximum heat-day practices

1. Increase water infiltration with dethatching or hollow-core aerification of all lawn areas, as well as under the drip line of trees. Till garden areas to break up surface crusting, adding mulch where appropriate.

2. Trim or prune trees, shrubs and bushes to remove low-hanging, broken or diseased parts and allow greater sunlight penetration throughout and beneath the plant. Generally speaking, plant water use is proportional to total leaf surface, thus properly pruned plants should require less water.

3. Fertilize all plants (when soil temperatures reach at least 50° F or 10° C) with a balanced plant food that contains nitrogen (N), phosphorus (P) and potassium (K) according to the results of soil testing or as experience has shown is appropriate.

4. Sharpen pruning shears and mower blades as dull blades encourage plant water losses and the introduction of disease.

5. Establish or confirm soil type(s) to match water-infiltration rates with future water-application rates and to determine if soil pH adjustments are recommended.

Locate drip emitters to provide water and nutrients needed by large trees, shrubs and other plants.

6. Perform irrigation-system maintenance, regardless of type (hose-end, drip, in-ground, etc.) to ensure maximum uniformity of coverage and overall operation. Repair or replace broken or damaged nozzles or heads. Flush drip system emitters to ensure proper flow. Ensure rainfall shutoffs and other devices are working properly.

a. Acquire and/or install hose-end water timers for all hose bibs.

b. Adjust in-ground system controllers according to plant's seasonal needs.

7. Upgrade in-ground irrigation systems by adding soil-moisture meters, rain shutoff devices or evapotranspiration (ET)-based controllers.

8. Relocate drip emitters, particularly around trees to the outer edge of their drip lines. While this will result in higher water use, it will also encourage a better root system that will anchor the tree in high winds and provide the water and nutrients that are needed by large trees, shrubs and other plants.

9. Confirm water-application rates for hose-end or automatic systems to know what actual running times are required to distribute a specified amount of water within a given amount of time.

10. Water in early morning when wind and heat are lowest to maximize the availability of the water to the selected plants.

Water in the early morning.

11. Irrigate all plants infrequently and deeply according to local ET or soil-moisture requirements to establish a deep, healthy root system. A core-extracting soil probe or even a simple screwdriver can help determine when to water if the more sophisticated ET rates are not available. Professional turf managers gradually lengthen the interval between irrigations to create gradual water stress for deeper rooting.

12. Cycle irrigation applications (on-off-on-off) to allow penetration and avoid runoff. Depending on soil types, the running times may be from 5 minutes to 15 minutes and off times from 1 hour to 3 hours. Repeat this cycle until necessary amounts of water are applied and maximum penetration is achieved.

An example of a core-extracting soil probe.

13. Adjust automatic timers of in-ground irrigation systems according to the plants' seasonally changing water needs.

14. Begin regular mowing when grass blades are one-third higher than desired post-mowing length, and keep clippings on the lawn.

15. Raise mowing height as summer progresses to the highest acceptable level to encourage deep rooting. (Note: While this has been a traditional recommendation, further study is required to refine this approach and maximize effective water use and/or conservation.)

B. Drought or maximum heat-day practices—to maximize landscape appearance

1. Withhold fertilizers, particularly nitrogen, on turfgrass; however, small amounts of potassium will aid in developing more efficient roots.

2. Reduce mowing frequency to minimize shock to turf areas.

3. Reduce traffic on turf areas as this will minimize wear and possible soil compaction.

4. Adjust automatic timers of in-ground irrigation systems according to the plant's seasonally changing water needs.

C. Drought or maximum heat-day practices—if dormant turf appearance is acceptable

1. Eliminate all traffic on turf areas including mowing, which will probably not be necessary because of the extremely slow growth rate.

2. Adjust automatic timers to manual or use hose-end sprinklers to apply approximately one-quarter inch of water a week. The dormant lawn will have a tan, golden or light brown appearance; however, light/infrequent watering will be sufficient to maintain life in the crowns of the grass plant during this period.

3. Minimize water applications for all plant materials to the essential amounts needed to maintain plant vitality.

A wasteful practice seen all too often is misapplication of water, resulting in flooded sidewalks or driveways and rivers of water wastefully flowing down gutters.

D. Post-drought or maximum heat-day practices—Irrigate all plants to re-establish soil-moisture levels, beginning with staged increases in watering to progress toward a deep and infrequent watering practice. By gradually lengthening the running times but adding greater spacing between watering applications, the initially shallower roots will extend to reach deeper soil moisture. Previously dormant turf will recover rather quickly, and other plants will regain their vigor.

II. Newly Planned or Installed Landscape Areas

While fewer in number than existing landscape areas, newly planned and installed landscape areas can generally achieve greater water savings if all of the best design, plant selection, installation and management practices now available are closely observed and fully implemented.

The basic principles of xeriscape landscaping provide an excellent starting point, providing they are fully understood and properly applied. However, it should be noted that it is incorrect to assume that a xeriscape is supposed to be a totally grassless landscape or one that uses only rocks, cactus or driftwood.

Dr. Doug Welsh, former director of the National Xeriscape Council, wrote in *Xeriscape Gardening: Water Conservation for the American Landscape:* "In xeriscape landscaping we try to plan the amount of turf so that the investment in water will be repaid in use and beauty. In many instances grass *is* the best choice. For play areas, playing fields and areas for small pets, grass is often the only ground cover that will stand up to the wear. Turf also provides unity and simplicity when used as a design component."

Rather than duplicate information contained in several xeriscape manuals, this publication will focus on a limited number of very critical elements that can maximize landscape water conservation.

Efficient irrigation is without question an important water-conservation activity. People waste water; plants don't. Overwatering not only wastes water, but it also weakens or kills plants more than underwatering. Another wasteful practice seen all too often is misapplication of water, resulting in rotted fences and house siding, flooded sidewalks or driveways and rivers of water wastefully flowing down gutters.

While less so today, many new in-ground landscape-irrigation systems have been sold in the recent past on the basis of simplicity, e.g., "set it and forget it." Homeowners, intimidated by the sophisticated appearance of the system's control box, would not modify the settings for seasonal plant water-use changes. Even worse, in order to be cost-competitive, many systems did not include readily available and relatively inexpensive soil-moisture meters, rain shutoff devices or multi-station programs. These deficiencies have resulted in overwatered landscapes, with water running down the streets and systems continuing to operate during torrential downpours.

Because of its high visibility, turf watering can be seen as the antithesis of water conservation and is often an out-and-out target for elimination. In some locales, "cash-for-grass" programs are used to pay homeowners handsomely to remove grass from their landscapes. One highly respected West Coast water official noted at a conservation convention: "It isn't the grass that causes a problem, it's the poorly designed and poorly operated irrigation system. I can't control the irrigation systems, but I can reduce the amount of grass in a landscape, and that will control the water-use problem created by bad irrigation." The fault was not the grass, but the fact that it was being improperly watered.

"It isn't the grass that causes a problem, it's the poorly designed and poorly operated irrigation system...."

Soil analysis and improvements is another very important aspect of water conservation. The soil on most new residential and commercial landscape sites have literally been turned upside-down during the construction process, with the topsoil placed beneath a layer of clay. Then it is compacted as hard as cement by equipment, piles of building materials and construction-worker foot traffic. The soil's texture, chemistry and natural flora and fauna are destroyed.

More water could be conserved and healthy landscape plants more easily grown by improving the soil before planting than by any other process or technique. While the initial cost of adding topsoil and soil test-guided amendments to improve the soil may seem high, the return on that investment will be even higher. Failing to improve the soil prior to planting, when it is most practical and efficient, will result in roots not being able to penetrate as deeply as possible and runoff occurring almost instantaneously. If the soil pH is not correct, plants will not be healthy, nutrients will not be utilized and chemical leaching can take place.

Appropriate plant selection can be a source of frustration or misunderstanding and not produce the hoped-for water savings, particularly when it is combined with "practical turf areas". As noted in Chapter III, lists of low water-use landscape plants can cause confusion because they are based on the incorrect assumption that those plants capable of surviving in arid regions are low water users, when the plants typically are only drought-resistant.

The majority of turfgrass species and cultivars have been scientifically assessed for their evapotranspiration (water-use) rates and can be selected according to the needs of a specific climate. On the other hand, very few tree and shrub species and cultivars have undergone comparable quantitative water-use assessments. One stunning exception comes from research at the University of Nevada, where Dr. Dale Devitt found that "one oak tree will require the same amount of irrigation as 1,800 square feet of low-nitrogen fertilized turfgrass!"

It is also important to understand that while "low water-use" plant lists were developed with the best

The mini-lysimeter gauges water use under actual turf conditions. This photo at Washington State University's research site in Puyallup, shows mini-lysimeters being weighed. The pots are weighed for daily water evaporation, rewetted and returned to their "holes" in the turf. These scientific measurements under controlled procedures help researchers better understand low-water use.

intentions and purposes in mind, practically all of these lists have been based on anecdotal evidence or consensus judgments, not scientific measurement under controlled procedures. Quite simply, a given group decides that based on their experiences and suppositions, a certain plant should or should not be placed on a "low water-use" list. As has been seen time after time, it typically is not the plant that wastes water, but the person who is in charge of its care.

When establishing new lawns, turfgrass sod has been shown to require less water than seeding.

When establishing new lawns, turfgrass sod has been shown to require less water than seeding, the evaporative losses from bare soil are greater for seeded areas than they are for turf-covered soil beneath sod.

Water "harvesting" and reuse is another water-conserving practice gaining greater use. It can be employed to conserve public water supplies and recharge groundwater supplies.

Historically, planners and designers have focused their efforts on moving rainwater and snowmelt away from a property as quickly as possible, giving little thought to the possible advantages of using that water for landscape purposes. Consider the fact that during a 1 inch rainfall, a 35-foot-by-60-foot roof (approximately 2,000 square feet) will collect nearly 1,250 gallons of water. Rushing this water to gutters and then sewers makes little sense when it could conceivably go into a system that could capture or distribute it across a landscaped area.

Another increasingly feasible source of additional landscape water is recycled or gray water. Some communities are installing dual water-delivery systems with one carrying potable water for drinking, cooking, cleaning and other general household uses and a second system delivering less thoroughly treated (but very safe) water for use on landscapes. On a small scale, some locales encourage collecting a home's gray water (from clothes washers, etc., but not toilets) for use on landscapes.

After being applied to a landscape area, harvested, recycled or gray water is either transpired by plants and evaporates into the atmosphere or finds its way into groundwater supplies after it has been cleansed by the plant's root structure.

As science and technology continue to advance, new and better information, tools and systems will become available to help people establish and maintain water-conserving and environmentally beneficial landscapes. An ongoing challenge will be keeping pace with these developments, sharing that information and continually improving best management practices.

Just as we improve our health by using water to brush our teeth and to wash our bodies and clothes, applying water judiciously to a properly designed and installed landscape can improve our health and the general environment.

Efforts to eliminate landscape water use not only take away freedom of personal choice; they also bring many environmental, economic and emotional drawbacks that could be more costly in the long run.

Chapter VII

Conservation-Aware Individuals Will Make the Difference

Summary:

An educated individual can be the start of a successful water-conservation program's "ripple effects." People will make sacrifices when they perceive there is a real need to do so. Conservation education programs should allow for personal choice.

Read between the lines of miles of flow charts and mountains of scientific reports on the subject of water use and conservation, and the answer to maximizing water conservation becomes evident.

It is the number "1."
One person...

- fully informed about the importance of water conservation
- adequately equipped with scientifically sound, yet simple principles
- properly motivated by reality, fear, costs or some combination of these

... can be the ultimate water conservationist.

One person can be the ultimate water conservationist

Those people who go about their everyday lives aware that their individual water use has an impact on their neighbors and everyone else "downstream" can make a difference. Most people can be persuaded to conserve for the greater good, but only when a real and proven need has been demonstrated.

Education Helps People Become Part of the Solution

Education and a reasonable plan that allows for personal choice are critical when individuals are called on to change old habits. Each of these elements deserves consideration and elaboration.

Rather than identifying people as being part of the problem, education can help people become part of the solution. While the content of any educational program on landscape water conservation must be tailored to fit local needs, some basic principles can be applied universally. These include:

1. Provide "early warnings" that alert people to a potential water shortage.

2. Provide clear, concise details of the depth and degree of the possible shortage.

3. Explain the background behind the potential shortage.

4. Suggest possible solutions, along with timelines and costs for each.

5. Give the public opportunities to participate and plan in prospective solutions.

6. Encourage new ways of thinking and practice regarding water use.

When people are asked to change consumption habits — especially if they regard the changes as a sacrifice — reasonableness is paramount. Members of a community should be asked to make the same degree of change. Those already conserving water should not be pressed further; those who give up wasteful patterns should not expect to be rewarded beyond knowing that they are doing the right thing.

Allowing for Significant Latitude of Personal Choice

But essential to the success of any effort to modify public attitude and actions must be a plan that allows for significant latitude of personal choice. When it comes to water conservation, the operative principle is that people should be informed as to *how much* water they can use but not *how* they have to use it. By maintaining their right of personal choice, individuals will be able to define their own needs and respond better than if they perceive the needs or plans of other people are being imposed on them.

Top photo: Although homes may sell with extreme xeriscapes (referred to by some residents as "desert landscapes") within a few years these designs are often modified with plants that reflected the homeowners' individual tastes (lower photo).

Demands and dictates, even when they are shrouded in a cloak of regulation or ordinance, have typically failed — particularly when it comes to as personal a matter as what one does at home. In the late 1980s planners of a new upscale residential development in Denver decided that xeriscape would be incorporated into all of the individual and common-area landscapes. Although the homes sold quickly, within a few years little of the xeriscape design remained because homeowners replaced it with plants that reflected their individual tastes.

In the year 2000, an "uprising" of sorts occurred in a gated Arizona community when its professional managers proposed removing nearly all of the traditional landscaped areas, including water-retention areas that have grass, to save on water and other costs. Tempers and voices rose when homeowners and managers met.

Essential to the success of any effort to modify public attitude and actions must be a plan that allows for significant latitude of personal choice. The operative principle is that people should be informed as to how much water they can use but not how they have to use it.

Protesting the proposed change, one homeowner said: "We will no longer have the beautiful views that were instrumental in our selection of this village and our specific property. Homeowners who paid a premium price of several thousand of dollars for their lots may find the property values could drop. Also, the current ecosystems within the larger basins may be changed to include rats, snakes and an increase in insect populations because of the desert landscaping."

The result of these proposals: rather than change the landscape, the community changed its watering practices to achieve water savings.

The Role of Water Meters

But even allowing for personal choice, people also need to understand the level of conservation that is required, and water meters are an essential component of any conservation program. Quite simply, if you don't know how much water is being used, you can never truly know how much (if any) is being conserved.

Based on measured water use, pricing can be a highly effective conservation motivator while providing for personal choice at the same time. Incentives to conserve or disincentives to waste can be built into water-pricing policies that can be easily modified according to the severity of a water crisis.

For example, a base rate could be established for what is determined to be a reasonable amount of water to be used for single-family dwellings or per unit for multi-family dwellings. With supply readily available, the price for additional units of water could be minimal. However, as a water crisis worsens, a series of water-use pricing tiers could be implemented. These pricing-by-volume tiers could be accompanied by a variety of multipliers that increase as the severity of the water crisis intensifies, thereby encouraging conservation over consumption.

Conservation can create massive "virtual reservoirs of water" wherever and whenever they are needed, but only to the extent that people can understand, accept and support the long-term value and benefit of making incremental changes that have substantial results.

Water is of little concern until the well starts to run dry. It is everyone's responsibility to ensure that doesn't come to pass.

Questions Water-Policy Officials Should Be Prepared To Answer

Ultimately, enlisting the public in a successful water-conservation program requires water-policy officials to be able to deliver on the promises they make — even if those promises are only implied. Officials should be prepared to answer:

Question 1. Are the landscape water-conservation goals short-term, long-term or indefinite, and are they expected to become more restrictive?

Question 2. Have the proposed conservation programs proven to be effective in saving water, or like some xeriscape programs actually resulted in using more water?

Question 3. Will the proposed landscape water-conservation efforts result in creating heat islands, which require increased energy consumption for cooling?

Question 4. Will there be any negative impacts on the environment and ecosystem as a result of the proposed water-conservation efforts?

Question 5. What are the economic impacts (positive and negative on home values, businesses and jobs) that can be expected as a result of these conservation efforts?

Question 6. What will happen to the water that is conserved? Where will it go and how will it be used?

Question 7. Will individuals maintain personal choice and individual responsibility when the proposed conservation efforts are implemented?

Case Study Contributors

Case Study 1: Never Underestimate the Importance of Plants to People

Dr. Paula Diane Relf is a professor of horticulture and an extension specialist in environmental horticulture at Virginia Polytechnic Institute and State University. She helped found the American Horticultural Therapy Association in 1973 and the People-Plant Council in 1990. A highly sought-after writer and speaker, her work has been recognized by the International and American Society for Horticultural Sciences. Dr. Relf received her doctoral degree from the University of Maryland where she pioneered studies in the area of horticultural therapy.

Case Study 2: 21st Century Landscape Water Use–A Global Perspective

Dr. James B. Beard is president of the International Sports Turf Institute Inc., College Station, Texas, and professor emeritus, Texas A&M University. In addition to being on the Texas A&M faculty for 22 years, he was a turfgrass professor at Michigan State University for 14 years. A renown researcher, speaker and author, he is widely recognized for his 1973 book, *Turfgrass: Science and Culture,* a leading college text. Dr. Beard earned his bachelor's degree from Ohio State University and his master's and doctoral degrees from Purdue University.

Case Study 3: Soil-Water Issues Relevant to Landscape Water Conservation

Dr. Eliot C. Roberts held faculty and administrative positions at The University of Rhode Island, University of Massachusetts, Iowa State University and The University of Florida. A ten-year appointment as executive director of The Lawn Institute closed out a career in agricultural science and education that has spanned some 50 years. Dr. Roberts holds Master of Science and Doctor of Philosophy degrees in Soil Science from Rutgers - The State University of New Jersey.

Case Study 4: Refining the Concept of Xeriscape

Dr. Douglas F. Welsh, is the campus-based extention horticulturist at Texas A&M University, College Station, TX, a position he has held since 1989. He is coordinator for the Texas Master Gardener program, a past president and board member of the National Xeriscape Council, Inc., and with over 17 years experience as a garden writer and broadcaster, co-authored the book, *Xeriscape Gardening: Water Conservation for the American Landscape*. Dr. Welsh was awarded his doctoral degree in horticulture from Texas A&M in 1989.

Case Study 5: No Water Should Be "Waste Water"

Ken Diehl is the water recycling specialist within the San Antonio Water System's (SAWS), Resource Development Department. Previously he was a senior aquatic toxicologist with ENRS Consulting & Engineering in Houston. Mr. Diehl received his bachelor of science degree in biology from Stephen F. Austin State University in Texas.

Case Study 6: The Important Role of Science in Landscape-Ordinance Development

Arthur J. Milberger is president of the Turfgrass America—Golf and Sports Division and Milberger Turfgrass, Bay City, TX. In addition to being a Lower Colorado River Authority board member, he was elected to the Turfgrass Producers International board of trustees in 1999. Mr. Milberger was awarded his juris doctor degree from St. Mary's Law School in San Antonio in 1974, having received a business honors program degree from University of Texas in 1971.

Case Study 7: Water Conservation on Golf Courses

James T. Snow is the National Director of the United States Golf Association's (USGA'S) Green Section, a position he has held since 1990. He is responsible for the Turf Advisory Service, Turfgrass and Environmental Research Program and the Construction Education Program. He is also editor of the Green Section Record, a bi-monthly magazine. He joined the Green Section in 1976. Mr. Snow was awarded a bachelor of science and masters degrees from Cornell University.

Case Study 8: Homeowners Can Conserve Water with Low-Tech and High-Tech Solutions

Thomas E. Ash is vice president of CTSI Corporation, Tustin, California, a water conservation service, products and implementation company. Previously he was the Irvine Ranch Water District conservation coordinator and a University of California cooperative extension specialist. Mr. Ash was awarded a bachelor of science in horticulture from California State Polytechnic University, Pomona, California.

Case Study 9: Maintaining Superior Landscapes on a Water Budget

Earl V. Slack is director of southern farming operations for Pacific Sod, based in Camarillo, CA, with experience in the turfgrass sod production industry since 1987. He served as the 2000 -01 president of Turfgrass Producers International. Mr. Slack was awarded an MBA from Pepperdine University in 1986, having received a bachelor's degree in agricultural science and management in 1979, from the University of California, Davis.

Case Study 10: Communicating Water Conservation to a Community

David Dunagan works in the Energy Efficiency and Renewal Energy Division of the U.S. Department of Energy, Atlanta, GA. He was with the Fulton County Environment Division's Water Quality Program. He serves on the board of directors of the Georgia Water Wise Council and the Southeast Land Preservation Trust. He was awarded a bachelor of science degree in forestry management from Mississippi State University and a master of science degree in public policy from Georgia Tech.

Case Study 1: Never Underestimate the Importance of Plants to People

Dr. Diane Relf, Professor of Horticulture, Virginia Polytechnic Institute and State University, Blacksburg, Virginia

"Intuitive arguments in favor of plants usually make little impression on financially pressed local or state governments or on developers concerned with the bottom line. Politicians, faced with urgent problems such as homelessness or drugs, may dismiss plants as unwarranted luxuries." Ulrich and Parson (1992)

Whether the debate is about the expenditure of a restricted number of dollars or the use of restricted amounts of water, our understanding and therefore our appreciation of the value that plants, landscapes and gardens have in our lives is severely limited. Too easily and too often, we approach the study and application of horticulture for human life quality with too narrow a focus.

A Broader Definition of Horticulture

If we broaden our definition of the relationship between plants and people, "horticulture" can embrace not only the art and science of growing flowers, fruits, vegetables, trees and shrubs. It can also result in the development of our minds and emotions, the enrichment and improved health of our communities and the integration of the garden in the breadth of modern civilization.

The Relationship Between Plants and People

Before off-handedly dismissing the value of investing money, water and other resources in landscapes, gardens, natural areas or simply plants and people, consider some highlights of the following exceptionally varied research findings:

- With a view of nature, recovery from stress was reported within 4-6 minutes, indicating that even brief visual contacts with plants, such as in urban tree plantings or office parks, might be valuable in restoration from mild, daily stress. (Ulrich and Simon, 1986)

Office park landscape

- Even in the urban environment with buildings, the presence of vegetation may produce greater restoration than settings without vegetation. (Honeyman, 1987)

A view of natural elements in the workplace

- Workers with a view of natural elements such as trees and flowers experienced less job pressure, were more satisfied with their jobs and reported fewer ailments and headaches than those who either had no outside view or could see only built elements from their windows. (Kaplan et. al., 1988)

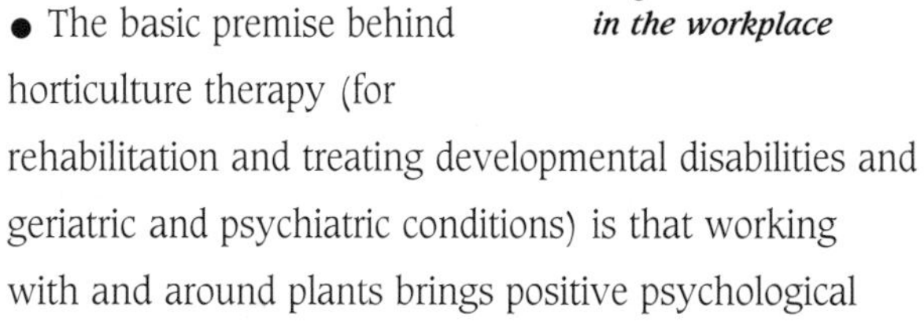

- The basic premise behind horticulture therapy (for rehabilitation and treating developmental disabilities and geriatric and psychiatric conditions) is that working with and around plants brings positive psychological and physical changes that improve the quality of life for the individual. (Relf)

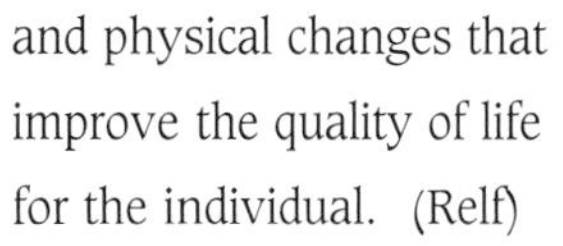

Horticulture therapy

- The physical condition of an area, be it a neighborhood or an office complex, provides a measure of self-worth of the area, defines

Landscaping can provide a sense of self-worth of an area.

the value of the individuals within that area and projects that definition to outsiders. If an area is dilapidated or vandalized, has trash-filled vacant lots or is sterile steel and concrete, it sends messages that those in charge (the city government, the owners, the employers) do not place value on the area and the people there. It implies that people have no intrinsic worth and no control over their environment and it tells outsiders: "This is not a good place to be." (Lewis)

- Partners for Livable Places maintains that plants are the fastest, most cost-effective agents for changing negative perceptions of an area, enhancing the economic and social conditions and improving psychosocial health. (Relf)

The landscape can influence residential property value.

- The strongest indicator of local residential satisfaction is the ease of access to nature, and this is the most important factor (after the marital role) to life satisfaction. (Fried, 1982)

- Parks and street trees were second only to education in the perceived value of municipal services offered. (Getz, 1982)

- The most important factors in neighborhood satisfaction among multiple-family housing complexes were the availability of trees, well-landscaped grounds and places for taking walks. (R. Kaplan, 1985)

- Residential property values are enhanced by their proximity to urban parks and greenbelts. (Correll and Knetson, 1978; Hammer et. al., 1974; Kitchen and Hendon, 1967)

- Professional appraisers estimated that unimproved residential land had a higher value if there were trees on the land, and a scattered arrangement was determined to have a higher value than concentrated arrangements of trees with the same percentage of tree coverage. (Payne and Strom, 1975)

Expanding Our Awareness of Our Environmental Relationship

The roles that plants play in social evolution reach far beyond food, fiber and medicine. The domestication of plants and animals allowed for massive changes in human culture. The act of cultivation brought intellectual, psychological and social rewards that are reflected in our folklore, literature and art. Gardens have been used as havens for reflection by philosophers and as sources of inspiration and symbols of virtue and vice by artists and poets. Plants and nature are woven into the unconscious human mind and serve as a source of spiritual renewal.

As horticulturists, water-policy decision-makers, politicians and citizens moving into the decade of the environment, we all will be called upon to expand our awareness of our relationship with the environment – and in the process we will enhance the value we place on the garden in the grand scheme of things.

Note: The preceding was excerpted from *HortTechnology* April-June 1992 2(2) and can be viewed in its entirety at http://www.vt.ecu/human/hihard.htm. Complete information on citations included in this excerpt is presented in the original document.

Case Study 2: 21st-Century Landscape Water Use: A Global Perspective

Dr. James B. Beard, Professor Emeritus, Texas A&M University; President, International Sports Turf Institute Inc., College Station, Texas

The extensive research I have conducted on the water-use rate and drought resistance of turfgrasses and the lecturing I have done more recently in 20 to 25 countries a year have given me a global perspective on landscape water use and its important role in various cultures around the world.

The Human Desire to Enhance the Living Environment

It is very significant that for 11 centuries, humans have chosen to devote time and resources, including water, to establish and maintain turfgrasses in landscapes for a better quality of life. While this desire to enhance the living environment may exist worldwide, it has not been attainable in those regions where people must spend all of their waking hours in pursuit of food, fiber and housing to survive. Countries that have industrial as well as agricultural employment can generate sufficient financial resources enabling individuals to afford to improve their living environment with landscape plants.

In my travels around the world I have consistently observed that countries with extensive urban landscapes, including lawns, trees and shrubs, also have associated with them a dominant population with a relatively high productivity rate. In addition, people in these places interact more harmoniously than people who live in areas that are seriously deficient in using landscapes to improve the quality of life.

Worldwide Landscape Water Use

I offer the following experiences and insights on landscape water use:

Northern Europe has been experiencing (at least on a short-term basis) a seemingly significant climate change to a more extended droughty period in the summer. As a result, governments have imposed water-use restrictions in Denmark, Luxembourg and parts of the United Kingdom. Extended drought stress is a new experience for landscape and turfgrass managers in northern Europe, as they typically depend on rainfall at fairly frequent intervals for most required water. Turfgrass managers have much to learn about the appropriate cultural practices needed to minimize drought stress. Drought also has brought out the inadequacy in all phases of existing turfgrass-irrigation systems.

Southern Europe, which is typically dominated by a Mediterranean climate including warm and dry summers, has historically used cool-season turfgrasses, based on practices common in England. However, current research and educational programs are attempting to introduce the use of warm-season turfgrasses as a water-conservation strategy. Italian researchers are leading these investigations and educational activities, with an emphasis on bermudagrasses (*Cynodon* species).

The use of bermudagrasses at the LeQuerce Golf Course in Nepi, Italy, is an example of using warm-season turfgrasses as a water conservative strategy.

In the **Middle East,** water shortages have been common for many centuries, and water quality is relatively poor because of high salt and/or sodium levels. Thus, emphasis has been placed on the use of salt-tolerant turfgrass species, such as seashore paspalum (*Paspalum vaginatum*).

Top photo: Native Paspalum growing on the Lanakai Beach, Oahu, Hawaii. Bottom photo: This salt-tolerant turfgrass species, seashore paspalum, is shown at the Sea Island Golf Course, Sea Island, Georgia.

A unique water-conservation strategy is being used in Israel on fine-textured soils that are 30 feet to 60 feet deep. Basically, this deep soil profile is recharged during the rainy winter season, when water costs are lower. Then, during the hot-dry summer period, deep-rooted dactylon bermuda turfgrasses are used to harvest the water, which is combined with a single irrigation per month to sustain green turf.

In **Africa** considerable attention is being given to the use of effluent water sources for landscape irrigation, as well as the use of salt-tolerant species such as bermudagrass and seashore paspalum.

In **Australia**, activists have been promoting legislation to minimize turfgrass areas and increase tree-planting programs. However, experiences in West Australia, which is the second-driest state in the world's driest continent, have proven the need for other considerations. Many decades ago, a pine plantation was established on the groundwater recharge and well field area that serves as a major potable water source for the city of Perth. The area is a very shallow sandy aquifer over impermeable clay. These trees have now grown to a substantial height and are actually causing an excessive drawdown of the aquifer due to the high evapotranspiration rates associated with the increasing canopy areas of the trees. Thus the Waters and Rivers Commission plans to conduct staged harvesting of the pine trees and plant these areas with a vegetative cover composed principally of low-growing perennial grasses that also will lend the areas a dual park-recreation function.

In **China** many decades ago during the Communist purges, the dictate was to eliminate symbols of capitalism throughout the country. Green lawns were removed and ornamental trees were cut down. Subsequently, many of China's outdoor public spaces have been maintained as well-swept, bare-dirt ground.

Some 15 years ago, I was contacted about the development of a revegetation plan for urban open spaces in cities such as Beijing. The elimination of green vegetative turf cover, which stabilizes the soil, had resulted in major atmospheric pollution in the form of dust storms. More importantly, the rate of serious human diseases was increasing much faster in Chinese cities than in other major cities of the world.

The Chinese had concluded that the lack of green vegetative cover and its associated living biological ecosystem of antagonists to disease-causing viral organisms had resulted in a major increase in these organisms, which were readily disseminated in wind-blown dust particles. Initial revegetation efforts emphasized tree planting, but this did not solve the problem. China is now embarking on an active program of revegetating open spaces with turfgrasses. These events illustrate the vital role of turfgrasses and the need for judicious water use to provide numerous functional benefits including the protection of human health.

Top photo: Tiannenmen Square in central Beijing, the site of the 1989 riots, was originally a solid gray mass of concrete. Bottom photo: in 1998, the Chinese government tried to soften the hard-line square by tearing up much of the cement and installing sodded bluegrass, giving it a more user-friendly appearance.

In the drier regions of **South America**, the primary problem is very archaic and relatively non-functional landscape-irrigation systems. A major investment is needed to improve these systems in order to achieve more efficient water use that in turn will maximize water conservation.

Singapore and **The Netherlands** are two outstanding examples of highly developed countries where the use of turfgrass and landscape plants is encouraged and people interact relatively harmoniously. While both of these countries are burdened with relatively dense populations and both have unique climatic and geographic situations, they are nonetheless two key examples of the importance and benefits of landscaping.

Parque Tezozomoc, Mexico City

Before: A 70-acre industrial area; After: the finished park for a community of one million people.

In one of Mexico City's most polluted areas, in the middle of an industrial and working-class district, was a space of 70 acres. This land was conceived as a cultural and recreational open space and transformed into a park for a community of one million people. The park was designed to recreate the topography and lagoons of the valley of Mexico as they were in the 15th century, to offer a symbolic vision of the region's historical and ecological evolution in an attractive and simple form. This physical memory is complemented with sports facilities, bike paths, a cafeteria, open-air auditorium and gymnasium.

The project was executed in four years, applying ecological concepts. The mounds were built of recycled earth from subway excavation; recycled water is used to fill the lake, for the irrigation and for year-round maintenance of the park; a municipal nursery was installed to produce plants for the reforestation of this part of the city. In the five years of its existence, the park has been transformed into a local landmark that receives from 5,000 to 20,000 visitors every weekend. This project has become the model to show that contemporary landscape design—even when done in the most difficult situations in terms of poor social conditions and extreme budgetary constraints—can provide a social, artistic and ecological benefit to a Third World environment such as Mexico City.

The architect was Sr. Mario Schjetnan / GDU and the project won the American Society of Landscape Architects President's Award of Excellence.

Case Study 3: Soil-Water Issues Relevant to Landscape Water Conservation

Dr. Eliot C. Roberts, Director, Rosehall Associates
Sparta, Tennessee

Because water and soil are the lifeblood of the planet, any policy that governs water use or proposes water conservation must incorporate the dynamic relationship that takes place between plants, the soil and water. The soil on which all plants grow is a highly valuable natural resource. We have a responsibility to protect it from erosion to conserve and enrich it for future generations, just as we have a responsibility to judiciously use our supplies of water and maintain or improve its quality. Landscape plants generally and turfgrasses specifically can help achieve both of these goals.

To realize these benefits, it's helpful to look at what scientists have learned about the characteristics of soil such as texture, size of soil particles, the presence of organisms, capacity to hold moisture, acidity-alkalinity and the presence of pollutants.

Soils are not homogenous, inert materials. They are composed of mineral particles that include sand, silt and clay, as well as living and dead micro/macro flora and fauna, chemicals, air and water in various percentages.

Sizes of the Mineral Components of Soil

From largest to smallest, mineral components of soil consist of the following:

- Stones = 10 to 100 millimeters in diameter (25 per inch).
- Gravel = 2 to 10 millimeters
- Coarse sand = 0.2 to 2 millimeters
- Fine sand = 0.02 to 0.2 millimeter
- Silt = 0.002 to 0.02 millimeters
- Clay = smaller than 0.002 millimeter. Clay particles are so small that they are measured in microns (0.002 millimeter is equivalent to 2 microns)

In terms of comparative size, if we enlarge a clay particle to be the size of an apple, then a silt particle on the same scale would be the size of a limousine and a medium sand particle would be the size of the White House in Washington D.C. Because soil particles have relative sizes this small, there are many of them. For example, a pound of medium sand contains about 2.5 million particles, while a pound of silt contains more than 2.5 billion particles and a pound of clay contains over 40 trillion particles. On the basis of total particle surface, one pound of sand would account for 20 square feet; silt would present 220 square feet of surface, and clay would have 5,500 square feet of particle surface.

In terms of comparative size, if we enlarge a clay particle to be the size of an apple, then a silt particle on the same scale would be the size of a limousine and a medium sand particle would be the size of the White House in Washington D.C.

The Multiple Components in a Landscape Rootzone

Also present in most soils are high microbe counts within the rootzone. Often there are more than 900 billion for each pound of soil. In each 1,000-square foot surface to a 6-inch rootzone depth, there will be a total of about 45 quadrillion organisms. As these organisms complete their life cycle and die, they deposit into the soil some 10 pounds of nitrogen, 5 pounds of phosphorus, 2 pounds of potassium, a half-pound of calcium, a half-pound of manganese and one-third of a pound of sulfur for each 100 pounds of dead organisms on a dry-weight basis.

Soil microbiological processes also convert organic matter into humus. This is an ongoing reaction of great importance. Humus helps to form and stabilize soil aggregates that are essential for deep and extensive root growth. Humus also contributes to the process within the soil that holds and releases nutrients for plant growth.

In addition, many small animals known as soil fauna occupy the root zones of plants and contribute to the living nature of the soil. Depending upon soil conditions that are favorable for these macro-organisms, from 1 million to 2 million may be present for each 1,000 square feet of rootzone. The live weight of these organisms would range from 15 pounds to 30 pounds per 1,000 square feet.

Of course the water molecule associated with the soil is exceptionally small. One fluid ounce of water contains approximately 1,000,000,000,000,000,000,000,000 (24 zeros, or one trillion trillion) molecules.

Grass Plants Have a Tremendous Potential for Root Growth

Grasses fit right in with the sizes and numbers of soil particles found within this fascinating system. For example, there may be as many as 35 million individual grass plants per acre, or about 800,000 per 1,000 square feet. No other type of plant culture involves such crowding. Roots grow down into the soil and it is there that grass plants have a tremendous potential for root growth — up to 375 miles of roots from one plant and as many as 14 million individual roots that may have a total surface area of 2,500 square feet. Thus, root numbers and surfaces fit well within the very small spaces surrounding aggregated soil particles.

It is important to understand soil properties so we can

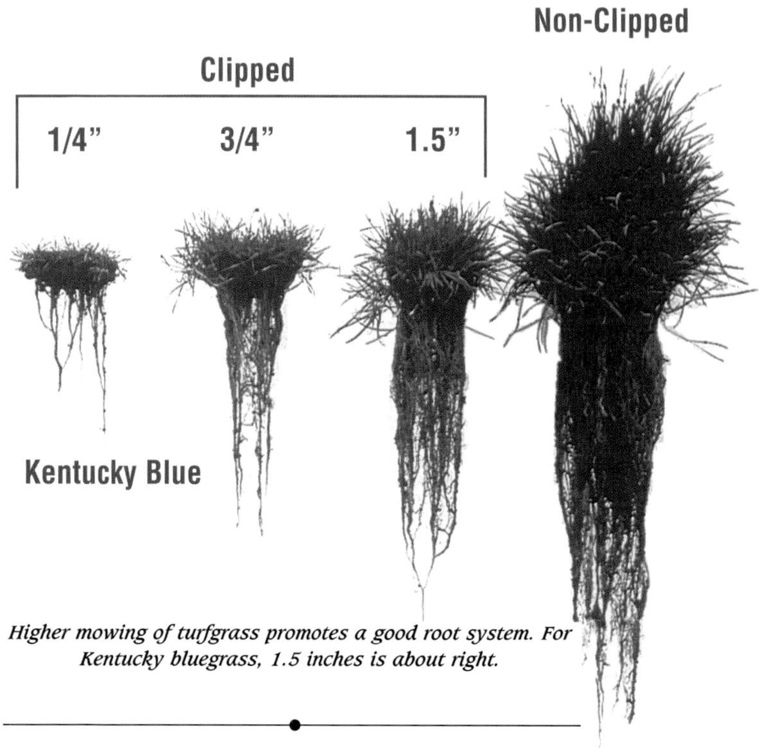

Higher mowing of turfgrass promotes a good root system. For Kentucky bluegrass, 1.5 inches is about right.

appreciate how turfgrass needs moisture in order to grow and enhance the environment.

Soil's Capacity to Hold Moisture

Soils differ in their capacity to hold moisture. Heavier clay and silt soils hold more moisture. Sandy soils can lose moisture through leaching as it runs through the rootzone and down into the subsoil. Grasses with well-developed, deep root systems add sufficient organic matter to help hold moisture in the soil and thus prevent leaching.

The Influence of Soil Texture on Water Penetration

The texture of the soil texture (as determined by the amount of sand, silt and clay) and the amount of thatch (organic deposit between green leaves and roots) influence the speed of water penetration into the soil. In general, heavy soils have many smaller pore spaces and take water in slowly. Sandy soils with fewer but larger pore spaces take moisture more rapidly unless they are inherently hydrophobic, or hard to wet. Soils and thatch that are hard to wet must be watered slowly with small amounts of water applied over longer periods of time in order to prevent runoff. Sandy soils require less water to penetrate to a given depth. Loam soils need intermediate amounts of water, and clay soils require more water to reach the same depth. As solid particles in the soil decrease on a percentage basis, moisture-holding capacity increases and soil aeration decreases.

The Degree of Acidity and Alkalinity in Soils

Soil may either be acid, neutral or alkaline. Soil pH (the measurement of the degree of acidity and alkalinity) is influenced by soil properties, biological influences and climatic influences. Under acid soil conditions, silt and clay particles tend to exist as individual units. Under more alkaline soil conditions, where calcium and magnesium are more plentiful, the clay and silt particles group together to form granules. These provide for improved soil structure, which results in more favorable balances of air and water in the soil. Where soils are acid and have poor structure, water penetration is much slower.

Soils become acid as carbon dioxide changes to carbonic acid in the soil, or acid-reacting fertilizers are used on a continuing basis, or acid rain-fall lowers soil pH values. Often a combination of all three of these causes occurs.

In addition to the effect of acidity on physical soil properties, nutrient fixation and availability also are modified depending on degree of acidity or alkalinity. For example, phosphates are most available from pH 5.5 to 7.5. Above and below these levels, phosphates are tied up with other minerals and their availability is reduced. Regular soil tests can determine need for lime or sulfur or for fertilizer mineral nutrients for specific plant types.

Soil as a Biodegradable Agent

Biologically healthy soil is the best-known medium for the decomposition of all sorts of organic compounds, including pesticides and pollutants transported by air and water. These chemicals are known to be biodegradable. This is an ongoing process, which changes these substances into harmless compounds plus carbon dioxide and water. Limited prescribed use of pesticides is not harmful to beneficial soil organisms and should continue to be an important, well-accepted part of plant culture.

Water and any pollutants associated with it infiltrate into the ground more quickly on grass-covered soils than any other surface. Thus, runoff is diminished. Infiltration rates may be as high as 7 inches an hour on sandy soils and as low as 0.10 inch per hour on clay soils. Thus, recharge of purified groundwater is an important benefit. An acre left in open space provides an average of 600,000 gallons of recharge each year in humid regions. Grasses may use up to 10 percent of the water infiltrated, leaving 90 percent for recharge of the local aquifer.

Dr. Thomas Watschke, Pennsylvania State University, created this highly controlled water-shed site documenting that established turfgrass has a dramatic, positive effect on reducing nutrient and pesticide pollutants from water runoff.

The Grass Groundcover Provides a Living Mulch

Good horticultural practice involves use of mulches to conserve soil moisture and increase soil productivity. Unlike many landscape plants that are either widely spaced, or simply annual in their growth habits, a grass groundcover provides a living mulch over the soil surface. This is essentially perennial and provides long lasting soil and water conservation benefits.

Instead of viewing green-lawn groundcovers as static liabilities, these areas can be seen as dynamic, ever-changing populations of plants and animals living within and above the soil. All grasses are natural soil builders. Particularly in residential areas, lawns and landscapes help sustain the soil. Within the soil are large populations of micro- and macro-organisms that are highly competitive. These create a living, moist soil environment best suited to sustaining productive landscape soils, while at the same time purifying our water supply.

Case Study 4: Refining the Concept of Xeriscape

Dr. Douglas F. Welsh, Professor and Extension Horticulturist, Texas A&M University, College Station, Texas

Much attention and controversy have surrounded the xeriscape concept of landscaping since its inception in 1981. The proper definition of xeriscape is "quality landscaping that conserves water and protects the environment." Above all things, it must be a quality design that balances the lawn area, shrub and flower plantings and the hardscape (i.e., decks, patios and sidewalks). Landscapes composed of rocks or plastic flowers alone are not xeriscapes. Xeriscapes are in tune with the environment; therefore, xeriscape applies to the desert southwestern United States as well as the semi-tropical southeast.

Xeriscape has seven basic principles:

- planning and design
- soil improvement
- appropriate plant selection
- practical turf areas
- efficient irrigation
- mulching
- appropriate maintenance

Use of Practical Turf Areas in a Xerixcape Design

Of the seven principles, none has received more attention than practical turf areas. This principle, which concerns turfgrass in the landscape, has been shrouded in misinformation that has been touted as fact by "experts" in xeriscape, water supply and turf culture.

The original turf-related principle established by the Denver originators of xeriscape was "limited turf use." For Denver and much of the arid West, the seemingly logical approach to reducing landscape water consumption was simply to reduce the use of turf. However, as the xeriscape concept has matured and spread, the principle of limited turf use was increasingly scrutinized by horticulturists and turf experts. Today's xeriscape movement incorporates a more holistic approach to reducing turf irrigation, fully recognizing that the type of plant materials or irrigation in the landscape has as much of an effect on water consumption as the human factor and good landscape water management.

The proper definition of xeriscape is "quality landscaping that conserves water and protects the environment." Above all things, it must be a quality design that balances the lawn area, shrub and flower plantings and the hardscape (i.e., decks, patios and sidewalks).

The Need to Change Attitudes and Habits

Throughout the xeriscape movement, the evident truth is that plants do not waste water; people do. Another fact is that irrigation systems do not waste or save water; people do. The mission of xeriscape is clear: Change the attitudes and irrigation habits of professional and amateur landscape managers. Proper water management provides the greatest opportunity for water conservation in the landscape.

Xeriscape focuses on the use of turfgrass in the landscape because of the tremendous potential for irrigation water abuses in the name of maintaining green turfgrass. Within the traditional landscape, turfgrass has received the major share of total landscape irrigation because grass often makes up a large percentage of the total landscape. Through the principles of xeriscape, turf irrigation can be reduced while the many benefits of turfgrass can still be derived.

Benefits of Turfgrass in the Landscape

Turfgrass is an integral component of most landscapes. It is certainly the best recreational surface for children and athletes. Furthermore, it has a tremendous mitigating effect on the environment, reducing heat loads, noise and water and air pollution. A turfgrass lawn is second only to a virgin forest in the ability to harvest water and recharge groundwater resources. As a design component, turfgrass provides the landscape with unity and simplicity while inviting participation in it.

However, the fact remains that turfgrass is the highest user of irrigation water in the traditional landscape. This is significantly different from saying that turfgrass is the highest water-using plant in the landscape – which is not the case. The discrepancy between these two statements yields the most common misconception and misrepresentation in xeriscape, and it is therefore the basis of controversy and unproductive efforts. To resolve this controversy, some scientific and practical fundamentals of turfgrass are explained using actual xeriscape principles.

Xeriscape Principles for Reducing Turfgrass Irrigation

Specifically, xeriscape principles promote the following strategies to reduce turfgrass irrigation:

- Prepare soils for turf areas as carefully as any other planting area to use all the moisture available, promoting the plant's vigor and water-use efficiency.
- Place turf species in landscape zones based on water requirements.
- Select adapted turf species and varieties that have lower water demands.
- Irrigate turf in areas that provide function (i.e., recreational, aesthetic, foot traffic, dust and noise abatement, glare reduction, temperature mitigation).
- Use non-irrigated turf areas where appropriate.
- Irrigate turf based on true water needs.
- Decrease fertilization rates and properly schedule fertilization.

Fine-tuning Turfgrass Xeriscape Principles

In traditional landscape design, turfgrass makes up the major portion of landscapes. The tremendous square footage of turfgrass in a landscape accounts for why turfgrass irrigation, as a percentage of total landscape irrigation, is so high. The "practical turf areas" guideline promotes the use of turf only in those areas of the landscape that provide function. In residential landscapes, a turf area is usually a necessity for recreation and entertainment. But turf should not be irrigated on narrow strips of land, or other areas that are difficult to water.

In residential landscapes, turf areas are important for recreation and entertainment.

Good landscape water management begins with planning and design. By designing the landscape as zones based on plant-water needs, turf can be appropriately placed for function, benefit and water efficiency. Zoning the landscape and irrigation system allows for watering turfgrass on a more frequent schedule than shrubs. For established trees and shrubs, the irrigation strategy should utilize deep soil moisture and depend on natural rainfall to replenish soil moisture. When sufficient rainfall does not occur, supplemental irrigation of trees and shrubs may be required.

Another way to incorporate turf into the landscape and conserve water is simply not to irrigate. Many turfgrass species are drought-tolerant and can survive extreme drought conditions. The grass may turn brown for a while, but rainfall will green it up again. This approach may be unacceptable for many residential and commercial landscapes, but in the case of parklands, industrial sites and rights-of-way, brown turf may be acceptable.

Selecting the Proper Turfgrass Species and Varieties

Wherever the landscape, selection of turfgrass species and varieties is of utmost importance. Extensive research has shown that there are significant differences in water requirements among turf species and even among varieties within species. The capacity of different turf species to avoid and resist drought also varies significantly. To help reduce landscape water requirements, xeriscape recommends selecting turfgrass varieties (and other landscape plants) that are both adapted to the area and have the lowest practical water requirements.

Landscape managers should be keenly aware of drought-stress indicators shown by turfgrass and other plants in the landscape, including a range of color changes, leaf curl and wilting, and they should strive to meet the water needs of each group of plants. By irrigating only when the plants require water versus by the calendar, the manager can dramatically reduce landscape water use.

Through specific horticultural practices, the water requirements of turfgrasses can be minimized. Decreasing fertilizer application rates and timely applications of slow-release fertilizers tend to reduce flushes of growth that can increase water requirements.

Attitudes and Habits About Turf Are Changing

For much of the arid West, the seemingly logical approach to reducing landscape water consumption was simply to reduce the use of turf. However, as the xeriscape concept has matured and spread, the principle of limited turf use was increasingly scrutinized by horticulturists and turf experts. Today's xeriscape movement incorporates a more holistic approach to reducing turf irrigation, fully recognizing that the type of plant materials or irrigation in the landscape has as much of an effect on water consumption as the human factor and good landscape water management.

Wherever the landscape, selection of turfgrass species and varieties is of utmost importance. Extensive research has shown that there are significant differences in water requirements among turf species and even among varieties within species.

The Xeriscape Challenge

Xeriscape is a challenge and an opportunity for the "green" (landscape, turf and nursery) and "blue" (water utilities and agencies) industries. Through xeriscape, these two industries have been brought together to focus on landscape water use. Although this marriage has not always been easy, the best minds are prevailing in efforts to perfect and implement the xeriscape concept.

By embracing the xeriscape concept, including the principle of practical turf areas, the green and blue industries can continue to be recognized as good stewards of the environment.

Case Study 5: No Water Should Be "Waste Water"— Fully Developing a Vital Water Resource

Kenneth Diehl, Water Recycling Specialist, Resource Development Department, San Antonio Water System, Texas

It is impossible to overstate the importance of groundwater supplies to San Antonio, Texas. The Edwards Aquifer is the sole source of drinking water for San Antonio, a city whose population has increased approximately 20 percent in the last 10 years to 1.1 million. The San Antonio Water System (SAWS) supplies an average of 170,000 acre feet a year to its customers. The Edwards Aquifer has reached a point where demands for pumping and springflows of water from the aquifer cannot be met from historical recharge by the underground sources that collect water to refill it.

Water Conservation Programs in San Antonio, Texas

Using existing water resources wisely, enhancing the Edwards Aquifer and developing new water resources are all critical to the continued progress and prosperity of San Antonio and the Edwards region.

San Antonio is situated in a part of Texas that receives only 29 inches of rainfall per year. SAWS is currently working on numerous water-supply resource projects such as aquifer enhancement, surface water availability, aquifer storage and recovery, and obtaining supplemental water from surrounding aquifers. These projects are at different phases of completion.

Using a Water Recycling Program Saves Drinking Water

The most cost-feasible of new water-supply projects, however, is to use the water that is already available — treated wastewater (recycled water). By utilizing a non-drinking water source for uses that do not require drinking-water quality, SAWS can supply these uses with recycled water, therefore saving drinking water for potable uses. SAWS has looked beyond traditional water resources to identify ways and means to economically, efficiently and effectively meet its current and future residential and commercial water needs.

After an exhaustive conceptual planning study, SAWS determined that it would be feasible to substitute 20 percent of its demands on the Edwards Aquifer with recycled water. The SAWS Board of Trustees approved the Water Recycling Program in 1996. The program treats wastewater to remove solids and bacteria while

When completed, the San Antonio Water System (SAWS) will supply San Antonio parks, golf courses and industrial customers with 11.4 billion gallons of non-drinking water each year. Photo: the Fort Sam Houston National Cemetery is participating in the program.

not bringing it to the level of drinking-water quality. The recycled water is then distributed (through a system separate from that which delivers drinking water) to non-residential users.

When completed, the program will supply San Antonio parks, golf courses and industrial customers with 11.4 billion gallons of non-drinking water each year, freeing enough Edwards drinking water for up to 80,000 families. The total cost of the project is $125 million, and completion of Phase I is scheduled for January 2001. This supply of non-drinking water will preserve drinking water and allow San Antonio the continued quality of life everyone has come to expect.

The Beginnings of the San Antonil Recycling System

Prior to the Water Recycling Program, about 120 million gallons of wastewater ran through approximately 4,300 miles of sewer lines in San Antonio every day. After treatment at four centers, the wastewater was released into the San Antonio and Medina rivers.

As part of the Water Recycling Program, SAWS created the separate recycled-water delivery system with approximately 73 miles of transmission trunk lines for Phase I, covering a 400-square-mile service area. At its current level, the system recycles about 11.4 billion gallons (35,000 acre feet) of water each year — and even with only a modest public outreach program, it has received requests for approximately 47,000 acre feet from 78 potential recycled-water customers. These include industrial and commercial concerns, agricultural water users, golf courses and parks. SAWS is already planning to expand the system in order to make more recycled water available in the near future.

SAWS' recycled-water customers understand that the treated water has the same cost and clarity as water that comes from the Edwards Aquifer and that its nutrient value is beneficial for landscape watering and most industrial applications. They also appreciate that this is an uninterrupted water source that is not affected by

The San Antonio Water System (SAWS) has created a separate recycling water delivery system with approximately 73 miles of transmission trunk lines.

drought restrictions, a factor that is particularly important to landscape-oriented businesses.

Bladerunner Turf Farms Is One of the Earliest Major Recycled-water Uses Within the SAWS System

One particularly progressive firm that illustrates the multiple benefits of recycled-water use is Bladerunner Turf Farms Inc, a Texas-based business that specializes in producing high-quality, drought-tolerant turfgrass sod for commercial and residential uses. Bladerunner was among the earliest major recycled-water users within the SAWS system, and it has become an important component in SAWS' ongoing testing and demonstration sites.

SAWS and Bladerunner currently have a formal agreement in which Bladerunner leases land from SAWS on two separate parcels adjacent to SAWS' Leon Creek and Salado water-recycling centers. Included in the long-term lease agreement is a provision for SAWS to provide 3 acre feet (977,500 gallons) of recycled water per year to irrigate the turf-production fields.

On the same site and working in conjunction with Bladerunner officials, SAWS established a very sophisticated turf-test and demonstration area next to the turf farmland. The purpose of the site is to scientifically evaluate the possibility of significant environmental effects that the use of recycled water may have on the land area that resupplies water to the Edwards Aquifer within the recharge zone. The site also will assist golf course main-

Kenneth Diehl, water recycling specialist for the San Antonio Water System (third from left) speaking to visiting turfgrass growers at the Bladerunner site where SAWS established a very sophisticated turf-test and demonstration area next to the turf farmland. The monitoring unit (left) is a solar powered weather station for determining evapotranspiration rates.

tenance personnel and San Antonio residents to learn about the proper application of recycled water to different types of turf.

The Turf Study Addresses Questions in Two Phases

With scientific instrumentation and procedures developed by a team of experts from Texas A&M's Soil and Science Department, the turf study will address questions in two phases. The first phase will focus on the possibility of nitrate contamination of the Edwards Aquifer from application of recycled water over the aquifer recharge zone. A secondary objective is to determine other agronomic best-management practices and to present a display site for Bexar County irrigators. The second phase of the study is designed to implement and evaluate the fate and mobility of fertilizer and pesticide.

Bladerunner will combine traditional turf-production practices with new knowledge gained from the turf study to help develop "real-world" practices that result in a quality product that is environmentally friendly and profitable to the farm owner.

Potential Customers Are Invited to Visit the Research Site

Ultimately, in addition to generating a substantial amount of scientific data for further analysis, the turf-study site will be open to individuals and groups so that they can gain invaluable knowledge about use of recycled water and advanced management of turfgrass.

By developing a system that effectively recycles 11.4 billion gallons of water a year, San Antonio has freed up the equivalent of 20 percent of the water that SAWS pumps from the Edwards Aquifer. Tapping into the recycling system is a way for commercial and industrial customers to help secure the water future of their community.

In addition to these important benefits and being able to use a large percentage of recycled water on landscape areas, San Antonio also maintains its reputation as a beautiful oasis while increasing the many environmental landscape benefits that accrue to the community, its residents and visitors.

The San Antonio Water System has found a win-win-win combination with recycled water.

Case Study 6: The Important Role of Science in Landscape-Ordinance Development

Arthur J. Milberger, President, Turfgrass America and Milberger Turf Farms, Bay City, Texas

"Don't Mess With Texas," isn't just a bumper-sticker slogan — it's an attitude that represents a way of life for the state's residents. Pride of ownership, independence, self-sufficiency and self-determination are all-important to Texans. Perhaps the most contentious topic among Texans isn't oil but water, and it has become even more important as the state's population continues to swell, but water resources in this arid southwestern state do not.

The General Services Commission of Texas Adopts Xerixcape Guidelines

It was against this backdrop at the end of 1993 that the Texas Legislature mandated the General Services Commission (GSC) of Texas to adopt guidelines for implementing xeriscape landscaping at all state facilities including buildings, roadsides and parks.
Incorporated into the state law was this definition of xeriscape:
" ... a landscaping method that maximizes the conservation of water by using site-appropriate plants and efficient water-use techniques. The term includes planning and design, appropriate choice of plants, soil analysis, soil improvement using compost, efficient and appropriate irrigation, practical use of turf, appropriate use of mulches and proper maintenance."

The Texas Legislature mandated the General Services Commission (GSC) of Texas to adopt guidelines for implementing xeriscape landscaping at all state facilities including buildings, roadsides and parks.

To adopt implementation of the guidelines statewide, the GSC was to consult with the Texas Natural Resource Conservation Commission, the Texas Department of Transportation and a newly created Industry Advisory Committee. The committee was composed of nine members: three nursery-product growers, three landscape contractors and three turfgrass sod producers. For better or worse, the GSC gave the newly named committee members only six days to respond to its proposed guidelines with additions, deletions and other comments. Turf-industry representatives felt the proposed guidelines had an unsupportable bias against turf in favor of other plants and trees.

Because the turf industry had addressed similar turf-restriction concerns in San Antonio and El Paso, its representatives to the committee understood that a war of words or conflicting opinions would not win over any turf opponents; however, unbiased scientific evidence could be a powerful tool. The representatives also respected their fellow committee members as hard-working volunteers trying to make a difference in their communities and the environment. The turf representatives also recognized that because this was a wide-ranging landscape ordinance, the concerns of the nurserymen and landscape contractors would be important.

Turfgrass Research Scientists Participate in Water Dialogue

Fortunately for Texans, scientists at Texas A&M and other universities in the state were investigating landscape water-conservation issues. Dr. Richard White and

Available scientific data shows that turfgrasses are not high water users compared with trees and shrubs.

Dr. Milt Engelke of A&M had recently consulted with the City of El Paso as it worked to develop and implement a landscape ordinance. White and Engelke wrote that their findings "suggested regulations should not seek to minimize or limit the use of any plant material in urban and suburban landscapes. Rather, efforts should be made to optimize all plant and non-plant material used in landscape designs to achieve the aesthetic appeal and functional qualities desired for a particular location and to ensure optimum and efficient use of resources required in the management of that landscape. Economics and water cost are already limiting the use of certain living plant material in landscapes. Educational programs and incentives for those who desire to use natural resources wisely, efficiently and the most economically in terms of impact on the environment and the pocketbook will likely have a greater impact than government regulation."

Dr. James Beard Responds to the Proposed Guidelines

In addition to scientists' statements that philosophically supported the use of turfgrass in xeriscape designs, specific and factually documented comments were presented regarding many elements of the proposed statewide guidelines. The following examples were presented by Dr. James Beard of Texas A&M.

Proposed guideline: Preservation of native plants that have been identified as desirable is encouraged.
Response: A native plant does not necessarily imply low water use or minimal maintenance.

Proposed guideline: Turf shall be used sparingly and only in circumstances where other landscaping media will not satisfy the site's needs.
Response: There is no known science-based justification for the statement. It is based on allegations and suppositions that turfs are high water users compared with trees and shrubs. The available scientific data show just the opposite. Water-use rates and transpiration [also known as "evapotranspiration" or "ET"] are associated with high-leaf canopy areas. Trees and shrubs have a much higher canopy area when irrigated than closely mowed turfgrass. The key is to select low water-use turfgrass, trees, shrubs, and flowers not exclusive of any one plant.

Proposed guideline: To reduce transpiration, extensive use of shade-producing trees is encouraged. However, only trees less than 30 feet high at maturity shall be used near overhead utility lines.
Response: This is an assumption that shade reduces evapotranspiration or radiant heat loads under the trees or surfaces (plants growing under them), and this is correct; however, the radiant heat load is transferred to the upper portion of the tree canopy, which has an extensive root system permeating the soil. This results in extensive use of water and a high transpiration rate from the upper canopy of many trees.

Proposed guideline: In planted areas, mulches of 2 inches or more shall cover most soil surfaces.
Response: Research has shown that mulches under trees reflect radiant energy onto the underside of the trees' canopy. This results in increased water use compared with the same trees [that have] bermudagrass turf growing beneath them. Mulches are rather expensive and result in a high maintenance cost relative to turfgrass due to erosion and weeding.

Research has shown that mulches under trees reflect radiant energy onto the underside of the trees' canopy. This results in increased water use compared with the same trees that have bermudagrass turf growing beneath them.

Proposed guideline: Maximum [allowable] percentages of turfgrass within an area shall vary, depending on the needs and uses of different types of facilities. Registered historical sites, cemeteries and athletic facilities are exempted from these percentages.
Response: A nebulous statement such as this should be avoided because it allows officials who are uninformed, possess a personal bias or are tied to particular industry to bring their biases into play.

Proposed guideline: Turf shall be irrigated separately from other plantings.
Response: This guideline apparently results from turf being associated with high water use, and it does not recognize the fact that trees and shrubs adjacent to turf use a significant percentage of the area's water.

Turfgrass Scientists Emphasize Focusing on the Big Picture

The scientific presentations emphasized that guidelines should not focus on a single issue such as turfgrass and that solutions should focus on the big picture, which includes water use, ground-water exchange, environmental protection, wildlife habitat and other concerns.

In 1994 the collaborative efforts of recognized turfgrass scientists, landscape contractors, nurserymen and land developers yielded *Xeriscape Guidelines for Texas State Facilities*, which also was viewed as a possible model for municipalities. The guidelines ultimately recognize the environmental benefits of turfgrass, and they were more "turf-friendly" than they otherwise may have been if turfgrass scientists hadn't been on the committee. The results, although they came out of a somewhat hurried process, must nevertheless be considered a success – the guidelines ultimately limited turf to 90 percent of landscaped areas.

The presentation of scientific facts enabled all participants in the process to gain a greater understanding of each other's concerns. Above all else, each constituent group was willing to keep an open mind about the others' ideas.

Case Study 7: Water Conservation on Golf Courses

James T. Snow, National Director, USGA Green Section, Far Hills, New Jersey

For several decades the golf industry has recognized its responsibility to reduce water use and become less reliant on potable irrigation sources. The industry has taken many steps to achieve this goal. Its multifaceted approach includes development of the following:

- new varieties of turfgrass that use less water or can tolerate poor-quality water
- new technologies that improve the efficiency of irrigation systems
- "best-management practices" in golf course maintenance that result in reduced water requirements
- alternative water sources that reduce or eliminate the use of potable water
- golf course design concepts that minimize the number of areas maintained with grasses that require considerable water use
- programs that educate golf course superintendents and other water users about opportunities for ongoing water conservation.

Improved Grasses that Require Less Water

Since 1982 the United States Golf Association has distributed more than $18 million through a university-grants program to investigate environmental issues related to the game of golf, with a special emphasis on the development of new grasses that use less water and require less pesticide. For example:

- Turfgrass breeders at the University of Nebraska have developed several improved cultivars of buffalograss (*Buchloe dactyloides*), which is native to the American Great Plains. This grass can replace high water-use grasses on fairways and roughs in a large area of the Midwest, resulting in water savings of 50 percent or more.
- Turfgrass breeders at Oklahoma State University have developed improved cold-tolerant, seeded-type bermudagrass (*Cynodon dactylon*) cultivars, allowing for the establishment of this stress-tolerant, low water-use grass in the transition zone of the United States to replace high water-use cool-season grasses. Water savings of 30 percent to 50 percent or more can be realized. When the Ruby Hill Golf Course in Pleasanton, California, was built several years ago, its fairways and

Turfgrass breeders are developing various grasses that thrive despite severe drought conditions.

roughs were established with bermudagrass instead of the cool-season grasses used at nearly all other courses in northern California. Ruby Hill estimates that it has a water savings of about 40 percent compared with similar courses that use cool-season grasses.

Catch cans determine irrigation distribution uniformity, after which adjustments will be made to improve uniformity and save water.

- Turfgrass breeders at the University of Georgia have developed improved cultivars of seashore paspalum (*Paspalum vaginatum*). This extremely salt-tolerant grass can be irrigated with high-salt or brackish waters with little negative effect on turf quality. Cultivars are available for greens, tees, fairways and roughs, and some of these varieties can be irrigated with water directly from the ocean.
- Ongoing breeding work is being undertaken on zoysiagrass (at Texas A&M), saltgrass (Colorado State and Arizona State universities), annual bluegrass (Minnesota and Penn State universities), alkaligrass (at Loft's, a seed company), fairway crested wheatgrass (Utah State University), colonial bentgrass (University of Rhode Island) and on a number of grass species at Rutgers University. This research, along with breeding work being done at other commercial seed companies, will provide new turf varieties for golf courses that reduce water use and pesticide use for decades to come.

New Irrigation-system Technologies

Tremendous strides have been taken in recent years to improve irrigation-system efficiency through the use of technology. These include:

- Using sophisticated on-site weather stations, weather-reporting services and other resources to determine accurate daily water-replacement needs, thus reducing the tendency toward over-irrigation. There also is a considerable effort being made to adapt various types of sensors to evaluate turf soil moisture-replacement needs, including tensiometers, porous blocks, heat-dissipation blocks, neutron probes and infrared thermometry.
- Improving irrigation uniformity through careful evaluation of sprinkler-head design, nozzle selection, head spacing, pipe size and pressure selection. The Center for Irrigation Technology at California State University, Fresno, is a leader in combining sprinkler uniformity and relative turfgrass quality needs to achieve the greatest water savings possible on golf courses and other turf areas. Many golf course irrigation-design companies and individual golf courses routinely use CIT's services to reduce water and energy consumption.
- Using state-of-the-art computerized control systems, portable hand-held controllers and variable frequency-drive pumping systems to apply water in the most efficient ways to reduce water and energy consumption.

This set of variable frequency drive pumps allows maximum flexibility in irrigating the entire course in the shortest period of time, improving efficiency and reducing energy cost.

These technologies can achieve considerable savings of water and energy resources. For example, the Southern California Golf Association Members Club in Murrieta recently installed a new state-of-the-

Best-management Practices for Golf Course Irrigation

Best-management practices for water conservation can be described as the combination of proper plant selection and horticultural-maintenance practices that provide adequate turf quality for the game of golf while minimizing water use. These include:

- Selecting low water-use turfgrasses, groundcovers, shrubs and trees for use on the course.
- Providing adequate levels of nutrients to the turf, including a balance of potassium and nitrogen, while avoiding excessive levels of nitrogen.
- Using mulches in shrub and flower beds to reduce water-evaporation losses.
- Adjusting mowing heights to ideal levels depending on species and seasonal water-use characteristics.
- Using soil-cultivation techniques such as spiking, slicing and core aerification to improve water infiltration and minimize runoff during irrigation or rainfall.
- Improving drainage where needed to produce a healthier turf with better root systems that can draw moisture from a larger volume of soil.
- Limiting cart traffic to paths in order to minimize turf wear and limit soil compaction.
- Cycling irrigation sessions to ensure good infiltration and minimize runoff.
- Pruning roots of trees near critical turf areas to prevent tree roots from competing with turf for moisture and nutrients.

art irrigation system that has reduced water use by about 35 percent. And because the club is able to complete its irrigation schedule in a short time frame during nighttime hours, it has reduced its considerable energy costs by about 50 percent.

Alternative Water Sources

It is not hard to understand why many communities are concerned about golf course use of potable water supplies, either from municipal sources or from on-site wells, during periods of drought and water-use restrictions. In response, many golf courses have developed alternative irrigation-water supplies and methods that do not depend on potable sources. These include:

- Storage ponds to collect storm runoff water that might otherwise be lost and wasted.
- Use of effluent that has undergone a three-step (tertiary) treatment process. This recycled water provides moisture and nutrients to the golf course while helping the municipality avoid discharging the effluent water into nearby rivers. The turf does an excellent job of filtering the water of nutrients and breaking down various chemicals and biological contaminants in the water. Use of recycled water on golf courses is mandatory in some locales in the Southwest, and it is estimated that more than 1,000 courses nationwide use recycled water.
- Use of brackish water or even ocean water to supplement other water sources. Bermudagrass is quite tolerant and seashore paspalum is very tolerant of high salt-content water, and these varieties allow golf courses to irrigate with brackish water that has few other uses. For example, the Old Collier Golf Club in Naples, Florida, is planting its greens, tees, fairways and roughs with two of the new seashore paspalum varieties emanating from the University of Georgia's turf-breeding program. The club will be using ocean water from a nearby estuary bay to irrigate the turf. A state-of-the-art irrigation system will allow precise water application for native plant materials, and the entire course will be irrigated during six off-peak hours to minimize energy costs.
- Building reverse-osmosis (RO) desalinization plants on the golf course to produce irrigation water from ocean water or brackish water where other supplies are not available or are very expensive. Three golf establishments in Florida and one in the U.S. Virgin Islands have built RO plants in recent years, establishing good-

quality, dependable and less costly supplies of irrigation water and allowing others in their communities to use the limited supply of potable water. They are the Everglades Club on the Barrier Island of Palm Beach, the Jupiter Island Club in Hobe Sound, the Sombrero Country Club in Marathon, all in Florida; and the Mahogany Run Golf Course, St. Thomas, Virgin Islands.

Golf course design concepts that save water

Today, golf course architects use innovative design concepts to help save water. These include:

- Careful earth shaping and good drainage design are used to collect runoff and sub-surface drainage water in on-site storage lakes.
- Turfed areas and water-demanding landscape areas are held to a minimum, resulting in water savings of 50 percent or more.
- Golf course sites with poor or inconsistent soils are capped with a 6-inch layer of sand to allow uniform water infiltration and a significant reduction in water use by reducing runoff and avoiding over-application of irrigation water.

A 12 million gallon retention pond is being built on this golf course to collect surface runoff and subsurface drainage water for supplemental irrigation..

The golf industry has taken many steps to reduce water use. This includes golf course design concepts that minimize the number of areas maintained with grasses.

Education Concerning Water Use and Conservation

- Numerous books related to golf course irrigation are available for irrigation practitioners.
- The Golf Course Superintendents Association of America and the Irrigation Association regularly present seminars concerning golf course irrigation.
- More than 2,000 golf courses participate in the Audubon Cooperative Sanctuary Program for Golf Courses, which educates course personnel about water conservation and protection and recognizes courses that take significant steps to conserve water.
- There are many industry periodicals that routinely explain and promote water-conserving practices.

These resources are likely to increase in the future as research continues and new technologies are developed. In addition, these newly developed and proven technologies and practices will be transferable to other managed turf areas such as sports fields, parks and home lawns.

Case Study 8: Homeowners Can Conserve Water with Low-Tech and High-Tech Solutions Alike

Tom Ash, Vice President, CTSI Corporation, Tustin, California

This case study will show how every home garden can be water-efficient, keep water bills low and reduce runoff.

Introduction

Home-landscape water use can consume up to from 59 percent to 67 percent of total home water demand (*Residential End Uses of Water*, American Water Works Association Research Foundation). What is the value of home landscapes, and how much water should a home landscape use? While there is no precise answer, attractive landscapes have been shown to increase property values from 7.28 percent to nearly 15 percent. How much water a home landscape needs depends upon its soil, sun and shade exposure, plant types, irrigation system and local climate.

To help meet current and future demand, public water agencies are seeking ways to gain verifiable, long-term efficiency in home landscape water use. The following examples show how simple and sophisticated tools alike can help public agencies and homeowners increase landscape water-use efficiency, save water, reduce peak demands and even manage periods of drought equitably throughout the community.

The techniques described below helped reduce home and commercial landscape water use in Irvine, California, by 50 percent from 1991 to 1999, saving consumers $28 million. (see Case Study 9).

Low-Tech Solution: Soil probes

"If we just had something that told us how wet or dry the soil is, we could save water," goes the saying of landscape managers and home gardeners alike. Such a tool does exist: the simple soil probe.

Horticulturists and university experts use soil probes to determine soil-moisture levels quickly. Using a soil probe is as easy as inserting it into the ground, pulling it out and then feeling and seeing the soil in it. To test the water-saving potential of this simple device, the Irvine Ranch Water District conducted a voluntary test of 90 homes. Residents were instructed to use the probes as follows:

1. Turn automatic sprinklers off.
2. Push the probe into the ground (where turf, shrubs, trees and groundcover are planted).
3. Pull out the probe and observe and feel the soil.
4. If the soil is wet or moist, do not water.
5. If the soil is dry, turn water on (using the probe to determine that water is applied only within the turf root zone).
6. Repeat the process before the next watering.

The first test group of 30 homes was monitored in the summer of 1997 (July-September). The second test group of 30 homes was monitored in the spring of

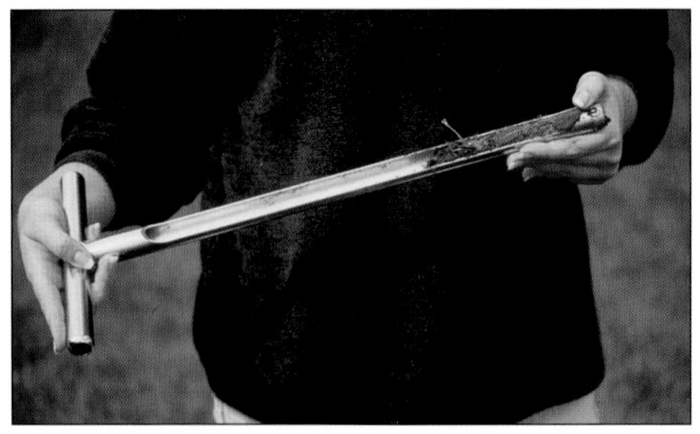

Using a soil probe for observing and feeling the soil for moistness.

1998 (April-June) and the third in the fall of 1999 (October-December). The test homes were compared with neighboring non-test homes (the control group) on the same streets. The water savings were measured against the previous year's water use for all test and control homes. The water savings for test homes over control-group homes were beyond agency expectations: 69 percent in the spring, 24 percent in the summer and 16 percent in the fall.

The $12 cost of the probe was recovered in average home water savings within each three-month test period regardless of the season. The cost-effectiveness of the probe combined with positive customer response makes it a simple and effective water-conservation tool — and homeowners and water agencies are using probes across the United States and around the world.

Soil probes are inexpensive and simple enough for everyone to use. They can save significant amounts of water regardless of the type of landscape and in the absence of water meters and sophisticated irrigation-scheduling technology. Water agencies, home builders and homeowners associations often give probes away as promotions at seasonal events and during home-water audits, and thousands of home gardeners are using the probes to reduce landscape water use.

High-Tech Solution: ET-signal irrigation controllers track weather and set efficient irrigation schedules

How much water plants require depends upon the type of plant and its evapotranspiration rate. Evapotranspiration, or ET, is the total amount of water lost from the soil through evaporation or used by plants to take in nutrients and control temperature. For healthy growth a plant needs only the amount of water the ET rate provides. Most plants suffer when they receive more water. Applying the right amount of water, based on the local weather and the plant's actual need, is the key to using water efficiently.

But gardeners often overwater, surpassing plants' real needs — and it is not difficult to understand why. Computing and setting landscape-irrigation time based on weather changes is a complicated, time-consuming and never-ending task. However, new irrigation-scheduling technology can change how water agencies and homeowners save landscape water.

The new wireless technology transmits local weather-station data each week directly to homes equipped with ET-receiving irrigation controllers, setting new and efficient irrigation schedules. This method of programming irrigation controllers provides the right amount of water at the right time for maximum plant health and water efficiency.

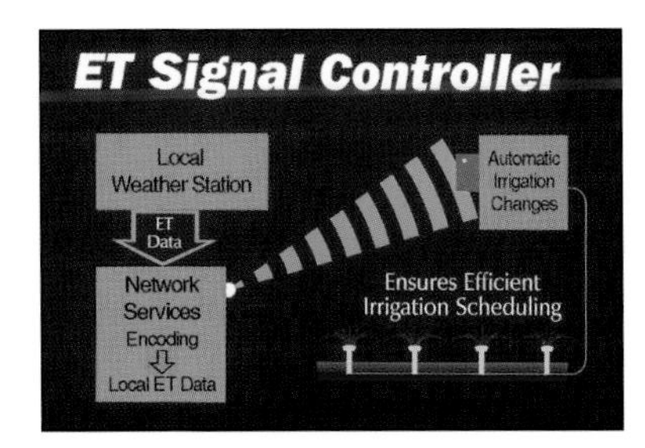

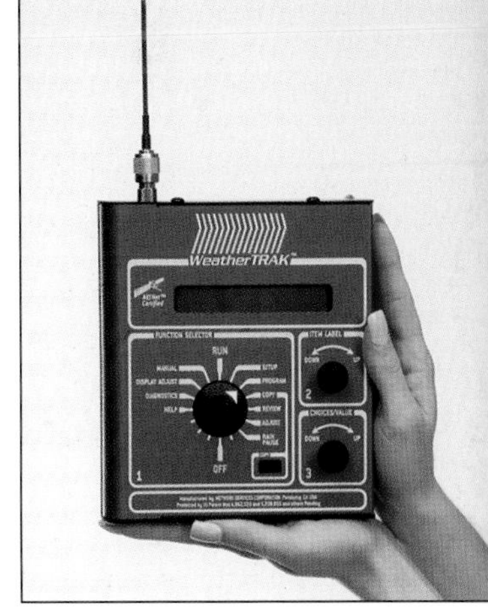

Above: The new wireless technology transmits local weather-station data each week directly to homes equipped with ET-receiving irrigation controllers. Right: The ET-signal irrigation controllers.

The technology was tested in a one-year study of 120 homes in Orange County, California, that was sponsored by the Metropolitan Water District, the Municipal Water District of Orange County and the Irvine Ranch Water District. The study evaluated the controllers' ability to perform three functions: set efficient irrigation schedules based on local weather, soil type, plant type, irrigation-system output and plant root depth; change irrigation schedules as the weather changes via a broadcast received by the controller; and eliminate the need for users to set, change or try to reprogram the controllers to meet plant requirements as the weather changes.

The study showed that the ET irrigation-control technology resulted in home-landscape water savings of 17 percent to 25 percent, and it indicated that water savings increase dramatically as the size of the landscape increases. One water-agency official observed: "For the first time in history we can achieve verifiable landscape water efficiency."

The test identified at least 10,000 homes in the Irvine Ranch Water District that could save water with the installation and use of this technology. But is it cost-effective for public agencies, cities and homeowners?

The study indicated that homes using moderate amounts of water for landscapes could save 57 gallons of water per day. This translates into an average annual savings of at least 20,000 gallons of water saved per home.

Irrigation-system controllers are a prime example of how technology and science can help policy-makers...

The study found these additional benefits of ET irrigation-control technology:

- All test-home residents said they found the controller to be convenient because they did not have to manually set, change or reprogram irrigation times.
- The average annual water cost-savings, at $114, was greater than the yearly ET weather-data broadcast signal fee of $48.
- Homeowners reported that their landscapes looked as good as or better than they did prior to use of the ET irrigation-control technology.
- The capability of the controller to be set for the soil-infiltration rate, slope and sprinkler output greatly reduces the potential for water runoff. (The U.S. Environmental Protection Agency is conducting an expanded test to project reductions of urban runoff from home landscapes using the ET irrigation-control technology.)
- The technology can be set to irrigate at specific or staggered times to assist with local supply shortages and/or occasions when local water use peaks.
- The technology can send a reduced-percentage signal prescribed during drought periods. (This need would be established by the local water agency to help meet drought and/or emergency shortages automatically. For example: If an area needs to reduce water use by 20 percent; the broadcast system can send an ET signal that is 20 percent lower across that part of the customer base that is outfitted with the technology. Landscapes can survive on 20 percent less water than ET standards, property values can be maintained and the agency can meet water reductions quickly and equitably.)

Further testing of this weather-based irrigation-scheduling technology will be conducted on different types of home and commercial landscapes in areas of the western United States that do not have water meters. The projection by the local water authorities is that ET irrigation-control technology can save water through continuous transmission of weather data regardless of the ability to measure home water use with meters.

The prospect of verifiable and efficient irrigation scheduling that saves landscape water has arrived. With the demand for water and the price of finding new water sources increasing, agencies and consumers are looking for simple, low-cost, credible solutions. ET irrigation-control technology has the demonstrated capability to save water and reduce urban water runoff from the landscape. It also can assist public water agencies in reducing peaking problems, and it can be used for drought management. Based upon the initial test results, agencies in the test area have begun to develop plans for large-scale consumer rebate and distribution programs.

With water demands projected to surpass delivery capacity in many areas, public agencies and landscape water managers will be able use ET irrigation-control technology to help meet water demands efficiently and cost-effectively. Irrigation-system controllers are a prime example of how technology and science can help policy-makers, planners, environmentalists, homeowners and businesses use water efficiently in urban landscapes.

Case Study 9: Maintaining Superior Landscapes on a Water Budget

Earl V. Slack, Director of Southern Farming Operation, Pacific Sod, Camarillo, California

Meeting today's urban water demands and assuring that future growth demands can be met have become increasingly difficult. Urban water agencies must consider a variety of approaches including but not limited to water conservation.

The Irvine Ranch Water District's Water Budget Program

Outdoor water conservation continues to be a major focus for many agencies because outdoor water use is highly visible. The Irvine Ranch Water District (Irvine, California) has adopted and successfully implemented a water-budget program that offers a viable and equitable solution to outdoor water conservation.

Water budgeting combines evapotranspiration-based irrigation scheduling and tiered pricing for increasing water usage. ET-based irrigation scheduling promotes conservation because it returns only the amount of water the plant needs and is based on well-established scientific principles. Tiered pricing provides an incentive to conserve because it gradually increases the price of larger quantities of water. These two tenets alone do not automatically translate into water savings, however. Education must go hand in hand with implementation so that consumers can use water budgets to maintaining quality landscapes.

Evaluating Three Interrelated Landscape Practices

To help document the combined conservation impact of these practices, a study undertaken in the Irvine (California) Spectrum Business District evaluated three interrelated landscape practices: evapotranspiration-based irrigation scheduling, improved irrigation-system maintenance and advanced horticultural turfgrass practices. Each of these components plays an important role in assuring that a quality landscape can be maintained within a water budget. Before the study was implemented, water usage at the test sites was more than 100 inches of applied water annually. The study found that implementing these practices resulted in a water-usage savings of slightly more than 50 percent, thus documenting the enormous potential for outdoor water-use reduction.

Evapotraspiration-based Irrigation Scheduling

The basis for water budgeting is evapotranspiration-based irrigation scheduling. The Irvine Ranch Water District based its schedule on 100 percent of cool-season turf's evapotranspiration (the total amount of water lost from soil by evaporation and used by plants to take in nutrients and control temperature).

Basing the water budget on 100 percent of ET ensured that the landscape received only the amount of water it actually needed. Before the study began, irrigation was often two or three times the rate of evapotranspiration. Thus, implementing evapotranspiration technology promoted water conservation.

The Irvine Ranch Water District evaluated turf "hot spots" to determine why they were happening. Landscapers, when they see such an area, increase water on that section and/or all sections on that controller. In reality, the "hot spot" was caused by lack of uniform water coverage due to clogged, blocked or sunken heads. The solution—repair the irrigation system.

Improved Maintenance of Irrigation Systems

Ensuring that the proper amount of water is applied depends on improved maintenance of irrigation systems. The study began by ensuring that the irrigation systems were functioning properly. System pressure was adjusted to avoid fogging and greater atmospheric loss due to high pressure and uneven water distribution due to low pressure. Sprinklers were aligned and set to proper elevation to allow even distribution. Leaks were detected and repaired to prevent direct water loss.

Using Advanced Horticultural Practices

Improved irrigation-system maintenance and operation allow for the most even distribution, but they also must be accompanied by advanced horticultural practices. The study identified the overall importance of soil quality in determining how well the landscape performs, and it documented that horticultural practices have two important objectives: improving the health and appearance of plants and increasing both root depth and the soil's capacity to hold water. These two objectives can be achieved through timing and selection of fertilizers and soil amendments as well as timing of irrigation, aerification and mowing.

The importance of deep roots should not be overlooked. Deep roots have a major impact on water conservation and the ability of turfgrass to grow well in dry weather. Soils have a modifiable water-holding capacity, and promoting deep rooting gives plants a much larger reservoir from which to draw. This allows irrigation frequency to be reduced. Improving the soil's water-holding capacity through aerification and amendments also helps to promote deeper rooting.

The Irvine Ranch Water District study determined that short roots, caused by watering practices, helped create the need for more water—which led to unhealthy turf.

Conclusions Based on the Irvine, California Water Programs

These three principles of evapotranspiration-based irrigation, improved system maintenance and advanced horticultural practices were combined to achieve significant water conservation without reducing the overall quality and appearance of the turfgrass. Study participants saw improvements in turf appearance and quality as the study progressed. This improvement occurred when the principles were implemented in combination but also when each principle was evaluated separately.

Every irrigation station received a detailed analysis including flow testing, precipitation rate, pressure and uniformity. The photo shows an irrigation station being measured with a "can test."

The study determined that tiered pricing and aggressive education accounted for a water-use reduction of 29.8 inches per year, and the advanced horticultural practices accounted for an additional reduction of 21.9 inches per year. Thus, this water-budget method produced a total savings of 51.7 inches per year, which is just over 50 percent of the pre-study water-use rate of 100 inches per year.

The ultimate conclusion: It is possible to conserve water using a water budget while maintaining a quality landscape.

Case Study 10: Communicating Water Conservation to a Community

J. David Dunagan, Energy Efficiency and Renewal Energy Division, U.S. Department of Energy, Atlanta, Georgia

July in Georgia is hot. It was hot in 1864 when General Sherman torched Atlanta before his infamous March to the Sea. The population of Atlanta, about 20,000 then, has grown to more than 3 million today, with the metro area sprawling across 17 counties. All the while, Atlanta, situated in the upper Chattahoochee River basin, has remained dependent on the smallest flow of surface waters to supply any city of its size in the United States.

The Making of a Crisis

The summer of 1988 was notably hot, and Georgia was suffering the cumulative effects of a rainfall deficit that had been building for three years. Because roughly one-half of peak summer water demand in the area is for outdoor water use, local governments began to impose water restrictions to conserve limited reserves. As summer progressed the restrictions were tightened until a total ban on outdoor water use was imposed. The inevitable protest that arose from the landscape industry and others who felt the ban's economic impacts fell on deaf ears in local government, which placed paramount importance on supplying drinking water and pressure for health and fire-suppression needs.

Water shortages are slow-building crises that do not capture the public's attention until a significant number of people are affected directly. That is when individuals and organizations become receptive to learning about conservation and about what they can do to help alleviate the crisis. The Georgia Water Wise Council was born in response to this 1988 crisis, when the issue was front-page news.

LESSON LEARNED: *Capitalize on these "teaching moments" of opportunity.*

Forming a Coalition

The key to building any successful organization is having the right players involved from the onset, each with an equal voice. In this case the Cooperative Extension Service of the University of Georgia took the lead. It began by holding ad hoc public committee meetings to form the Georgia Xeriscape Council. Attending were representatives from the extension service, the state's green industry, state and local governments and the water utilities. Consensus was reached that long-term public education for landscape water conservation was needed to avert future crises and to allow for future population and economic growth. As a result, in December 1989 the more-encompassing Georgia Water Wise Council (GWWC) was established and registered as a non-profit education corporation.

From its inception the GWWC maintained political neutrality, focusing on education goals while providing a forum for the exchange of ideas and information. Its members share an interest in water across political boundaries and economic sectors. This allows them to better appreciate each other's concerns. Invaluable informal bonds have been created and expertise freely shared.

And the members have learned from each other. Ten years ago, there was tension between the green industry and water producers due to a mutual lack of understanding. Now there is cooperation and accord. Fox McCarthy, a founding member of the GWWC and a former water-conservation coordinator for the Cobb-Marietta Water Authority, put it succinctly: "These utility guys pay attention to the green industry and conservation now.

...They're no longer old water buffaloes who just want to sell water."

LESSON LEARNED: *A balanced coalition yields a consistent message with minimal controversy.*

Materials in the Water Sourcebook series are in the public domain, so teachers can photocopy and distribute any part of them as needed.

Communication components

• **Water Sourcebooks:** The Water Sourcebook series is a set of four curriculum guides that are divided by grade-level ranges from kindergarten through high school. The books contain hands-on water education activities and science-lab demonstrations that are easy for teachers to present using readily available materials. These lesson plans are designed to enhance existing curricula through interdisciplinary teaching of mathematics, science, language arts and social studies. Correlation sheets guide teachers to water-education activities with the emphasis they feel their students need.

The Water Sourcebooks were funded primarily by the U.S. Environmental Protection Agency (Region 4) and developed with the Alabama University System Colleges of Education. The materials are in the public domain, so teachers can photocopy and distribute any part of them as needed. But it is less expensive and easier to buy the sets from the GWWC, which contracts for inexpensive printing and sells the sets for approximately $23 each.

The sourcebook material is being converted to CD-ROM format for easier access, storage and use. Distribution of Water Sourcebooks through the GWWC provides an added value: the availability of hands-on teacher-training workshops. These affordable sessions guide teachers through activities provided in the books. The experiments, games and demonstrations are applicable to all geographic areas — and they provide a public-education component to complement the education efforts of governments and public utilities that buy the books for teachers in their school systems.

LESSON LEARNED: *Make educational materials affordable and easy to use. Multiply your impact with "train-the-trainers" workshops.*

• **Print media:** Over the years the GWWC and its members have written dozens of articles for publication in newsletters and trade journals. Drawing on the specialized knowledge of council members, articles on topics including balanced landscapes, efficient irrigation and drought-tolerant cultivars are offered for publication at no cost. One well-written article can serve numerous newsletters. The Georgia Green Industry Association Journal ran a series of 28 articles by GWWC members in four issues.

LESSON LEARNED: *Use existing systems to spread your message.*

• **Trade associations:** Green industries have supported the GWWC's conservation-education efforts by providing free exhibit space at their annual conventions since 1990. In return, these industries can count on the council to provide speakers and articles to help inform industry members about water conservation. The arrangement is mutually beneficial and broadens the forum for conservation efforts. A video about water conservation for landscapers, recently produced by the Cooperative Extension Service in English and Spanish, can be played at conventions on a continuous loop. Booth space has been donated by the Southern Nursery Association, the Georgia Turf Association, the Georgia Green Industry Association and the Georgia Water and Pollution Control Association.

LESSON LEARNED: *Groups with related interests can help each other to achieve mutual goals.*

According to the Georgia Water Wise Council's guidelines, early xeriscape programs urged minimal turf usage, whereas current programs emphasize the strategic use of high-quality turf areas—"practical turf areas," for the most functional benefit.

• **Scholarships:** The GWWC established a $25,000 endowment for permanent funding of four 4-H Regional competitions that are judged by the Cooperative Extension Service. Students who make water quality or conservation a part of their project are eligible to compete for $500 scholarships. The council also recently established a similar program providing four annual $500 grants to teachers in order to support the use of Water Sourcebooks as the basis for conservation projects. The program is administered through the Georgia Science Teachers Association, and the GWWC is not involved in judging for any of its grants.

LESSON LEARNED: *Leverage your message exposure by creating well-publicized competitions.*

• **Xeriscape:** Teaching the seven principles of xeriscape – quality landscaping that protects the environment and conserves water — has been the centerpiece of the GWWC's landscape-education efforts. To complement those efforts, the Cooperative Extension Service produced a user-friendly 40-page reference guide, *Xeriscape: A Guide to Developing a Water-wise Landscape*. The guide, which has a companion video and scripted slide set, was published with sponsorship arranged by the GWWC, which then served as a primary distributor of these low-cost teaching materials. Any governmental or other organization wishing to promote water conservation through sound horticulture can easily adopt this ready-made material. All of the Atlanta-Fulton public libraries have received copies of the books and videos for use by the general public. In addition to these educational materials, the GWWC provides training and advice to organizations that want to start programs of their own.

The message comes to life at state-owned xeriscape demonstration gardens in Griffin and Savannah that were built in part with financial support from the GWWC, and in live presentations given by council members. Thousands of people have been exposed to the concept when visiting a xeriscape booth at the annual Southeast Flower Show.

Teaching the seven principles of xeriscape—quality landscaping that protects the environment and conserves water—has been the centerpiece of the GWWC's landscape-education efforts.

LESSON LEARNED: *Successful organizations support the initiatives of their members.*

• **Internships:** Water utilities that promote conservation can turn to the GWWC to connect them with interns from the University of Georgia School of Environmental Design. Once trained and supplied with materials, the landscape-architecture interns conduct audits using checklists, and they advise homeowners as a public service. The interns get credit toward their degrees as well as a rewarding experience and summer income. Cobb County, Georgia, runs a highly successful landscape-audit program.

LESSON LEARNED: *Interns can turn to their local cooperative extension agent for backing if they hit any snags.*

An example of rainwater harvesting using retention ponds. This photo is a scene from the Carter Center in Atlanta, Georgia, where the water is used to irrigate the landscape.

Pointers from the Georgia Water Wise Council Experience

• Cultivate in-kind services from member organizations to leverage funds.

• Keep your message clear, simple and apolitical. Make it easy to replicate.

• Learn from each other to refine the message. For example, early xeriscape programs urged minimal turf usage, whereas current programs emphasize the strategic use of high-quality turf areas, i.e. "practical turf areas," for the most functional benefit.

• Maintain continuity and retain organizational memory by mentoring newer members.

• Encourage involvement and contribution from all players to make everyone feel valuable.

• Encourage networking within the organization. Helping each other freely benefits all.

• Whenever possible, use existing systems such as neighborhood newsletters, trade associations, 4-H, etc. to spread your message. More exposure generates more requests for programs and information.

• Take advantage of moments of opportunity. For example, GWWC supported the creation of the Georgia State Xeriscape Demonstration Gardens.

• Establish your group as one that gets things done. Try new programs, then evaluate and refine them.

• Like any successful promotion, communicating about conservation requires a sustained effort.

Appendix A

Indoor & Outdoor Residential Water Conservation Checklist

There are many ways to conserve significant amounts of water inside and outside the home, and doing so makes sense because it lowers water and sewer bills, extends the water supply and helps the environment. Homeowners can take many simple steps to help preserve this precious renewal resource.

Your water-delivery system

Effective water conservation requires awareness, involvement and education. To understand your water-delivery system, know the following information:

☐ The name and location of the company that provides your water, as well as contact information for the company's chief executive and public education/public relations officials.

☐ Who the water-policy decision-makers are in your municipality or area, how they are selected (elected or appointed) and the length of their terms of office.

☐ How water-use policies and rates are set and modified, including names and contact information for officials.

☐ When and where announcements of public water-policy meetings are published (newspapers) or posted (office and/or Web sites).

☐ The source(s) of water used within the system (e.g. lakes, streams, groundwater or aquifer) and how to track stability and quality of supply.

☐ The water supplier's long-term and short-term contingency plans to ensure availability.

☐ The water supplier's contingency plans in case of supply shortage or interruption due to an act of nature (e.g. flood or drought) or mechanical failure of the piping, pumping or filtration system.

☐ The rate structure for residential, commercial or industrial water use, with possible seasonal modifications. (Note: Water-use billings may or may not include sewage-treatment fees, or they may be linked to potable water volume.)

☐ The location of the on-site water meter and how to read it and calculate the quantity of water used between readings.

Indoor water conservation

☐ Repair all water leaks immediately and be especially alert for leaks in toilets and faucets.

☐ Install and maintain ultra-low flow toilets. Alternatively, convert existing toilets to low-flow units with a tank dam or even bricks.

☐ Install and maintain flow restricters (aerators) on faucets.

☐ Install and maintain low-flow showerheads.

☐ Limit showering time to 5 minutes.

☐ Do not use toilets as waste baskets or ashtrays.

☐ Turn off water when shaving and brushing teeth.

☐ Scrape food off dishes without water prior to rinsing.

☐ Operate dishwasher only when it is fully loaded.

☐ Operate clothes washer only when it is loaded to maximum capacity.

☐ Rather than run the tap for cool drinking water, keep a filled container in the refrigerator.

☐ While waiting for running water to warm or cool for use on plants or in cleaning, capture flow for other uses.

Outdoor water conservation

☐ Cover pools, spas and other water features when not in use to minimize evaporation.

☐ Clean sidewalks, driveways and patios by sweeping rather than by spraying with a hose.

☐ Wash car(s) with a bucket of water rather than a running hose. If possible, drive your vehicle onto the lawn so that all of the water can be absorbed into the landscape.

☐ Restrict or eliminate use of hose-end water toys. If possible, combine use of water for play with landscape needs.

☐ Properly prune or trim trees, shrubs and other woody plants to maximize the plants' health and minimize invasion by pests.

☐ Frequently remove dead or dying plants and all weeds that compete for available water.

☐ Apply fertilizers or pesticides at minimal levels, timed to specific needs of the plants.

☐ Maintain sharp blades on pruning shears and lawn mowers.

☐ Aerate lawn and cultivate planting beds periodically to decrease compaction and improve penetration of water, air and nutrients into root zones.

☐ Mulch flower and garden areas as well as tree and shrub bases as appropriate for each species.

☐ "Harvest" water from rainfall and snowmelt for landscape irrigation purposes.

☐ Use recycled or non-potable water to the greatest extent possible, as limited by supply and/or regulation.

☐ Employ a certified landscape-irrigation auditor at least once every five years to conduct a thorough and comprehensive check for efficiency of water application.

☐ At least once a year, confirm that all irrigation systems are distributing water uniformly and inspect, repair and/or adjust inground or drip watering systems.

☐ Use water timers or flow meters for hose-end watering to ensure proper amounts are applied.

☐ Immediately shut off irrigation system(s) and adjust whenever irrigation water falls or runs onto hard surfaces such as sidewalks, streets or driveways.

☐ Repair all water leaks as soon as detected. (This includes leaking hose couplings, hose bib leaks and similar connections.)

☐ When buying plants, select those that have scientifically documented low water requirements.

☐ Determine specific water requirements for all existing landscape plants.

☐ Adjust controllers for in-ground or drip watering systems according to seasonal needs of plants.

☐ Water landscape plants only when necessary according to needs of each plant type or based on local ET (evapotranspiration) rates.

☐ Water early in the morning when temperatures and winds are at their lowest levels to reduce evaporation.

☐ Water all plants deeply but infrequently to encourage deeper, healthier rooting.

Appendix B

Landscape Water Conservation Ordinances

Public ordinances are intended to promote health, safety and general welfare. Ordinances that regulate landscape water conservation must take into account a wide variety of city- or site-specific considerations because what may be essential in one area may be impractical or dangerous in another.

The following major points are recommended for inclusion in all landscape water-conservation ordinances, subject to refinement as called for by location.

FINDINGS OF FACT

WHEREAS, the (city/county of ___________) recognizes the need to protect and preserve water as a natural resource through the application of enhanced landscape practices; and

WHEREAS, the (city/county of _______________) recognizes the need to encourage the quality of life, the freedom of choice and the emotional and economic values resulting from individually owned and public enhanced landscapes; and

WHEREAS, all residents of the (city/county of ______________) enjoy an unalienable right to artistic expression and personal choice within the bounds of public health, safety and general welfare; and the value of a given plant shall not be determined solely by its need for or consumption of water; and

WHEREAS, landscape water conservation reduces energy expenditures in the individual landscape, thereby lessening community energy expenditures for water pumping and treatment, and

WHEREAS, properly designed and maintained landscapes reduce urban heat islands and residential energy consumption required for air conditioning, and

WHEREAS, the (city/county of _______________) recognizes that there is no universal answer to all landscape water-management or conservation issues,

THEREFORE, landscape water conservation solutions shall be based on site-specific determinants, incorporating both initial establishment and continuous, long-term considerations.

PURPOSE AND INTENT

The purpose of these regulations is to establish minimum standards for the development, installation and maintenance of landscaped areas without inhibiting creative landscape design. Implementation will aid in improving environmental quality and the aesthetic appearance of public, commercial, industrial and residential areas. It is the intent of this ordinance, therefore, that the establishment of these minimum requirements and the encouragement of resourceful planning be incorporated to promote the public health, safety and general welfare in the areas of water conservation and quality preservation.

GENERAL PROVISIONS

A. Planning and Design

1. **Water budgets (allocations)** shall be established based on the area's climate and size of the property, with the maximum water allowance of 100 percent of the area's reference evapotranspiration (ET). Site owners shall have full and exclusive authority and responsibility to balance the design, installation and maintenance of their landscapes within this designated amount of water.
2. **Topography, grading and guttering** shall, to the maximum extent feasible, incorporate the concept of "water harvesting." This results in the greatest possible use by landscape plants of natural precipitation (rainfall or snowmelt) while minimizing the rapid movement (runoff) of this or other moisture into a stormwater drainage system.
3. **Fire protection** shall be addressed by giving preference to irrigated grass areas. The use of plants whose growth habits encourage or fuel fires shall be discouraged.

4. The use of **grassy buffers** shall be encouraged for lands adjacent to or contiguous with open waterways or known groundwater recharge areas. Such buffers slow erosion and cleanse runoff as it passes through the blades and dense, fine root structure of grass.
5. **Impermeable and covered surfaces** of not more than _____ percent (___%) of the total lot area may be incorporated into the design.
6. **Plant selection and grouping** choices shall include consideration for adaptability to climatic, geologic and topographic conditions of the site.
   a. Plants with the same known water-use rates may be grouped to facilitate water-use efficiency.
   b. Groupings of plants within the same drip line of large shrubs and trees should have water requirements similar to each other and the shrubs/trees because water will be shared by all of the plants in the group.
   c. There shall be no restrictions or limitations on the suitability of any type of landscape plants except those specifically prohibited by noxious-weed or invasive-species laws of this or a superior jurisdiction.
7. **Water features** (such as pools and spas), because of their high potential for evaporative water loss, shall utilize recirculating water exclusively.

   The year-round use of **pool and spa covers** shall be strongly encouraged.

B. Soil Testing and Modification

1. **Soil testing** shall be strongly encouraged to determine the type(s) of both the existing soil and the amendments that would be as favorable to landscape water conservation as possible.
2. Based upon test findings, **soil amendments** will be added to the site to the greatest extent possible prior to planting.

C. Irrigation System Design, Installation and Maintenance

1. **Soil types and infiltration rates** shall be given primary consideration when designing irrigation systems. All irrigation systems shall be designed, installed and maintained to avoid runoff, low head drainage, overspray or similar conditions in which water flows onto adjacent property, non-irrigated areas or hard surfaces such as walks, roadways, driveways and patios.
2. **Proper irrigation equipment and schedules** — including features such as repeat cycles, rain-sensing override devices, soil-moisture sensing devices and evapotranspiration (ET) rate-signaling controllers — shall be used to the maximum extent possible to match application rates to infiltration rates so runoff will be minimized.
3. **Outdoor water-use measurement** shall be strenuously encouraged through the use of separate meters, hose-end meters, timers or other accurate devices.
4. **Recycled water use** shall be encouraged to the greatest extent possible.
5. **The practice of water harvesting** shall be encouraged to the greatest extent possible.
6. **Landscape-irrigation audits** shall be required for all properties of one acre or larger at least once every five years.

(**Note:** *We would like to acknowledge that concepts and specific language have been extracted in full or in part from "A Water-Efficient Landscaping Guide for Local Governments, 2nd edition," prepared by the St. Johns River, Southwest Florida and South Florida Water Management Districts; and the "Model Water Efficient Landscape Ordinance" of the California Code of Regulations.)*

Principles of Efficient Landscape Water Management

Turfgrass Producers International (TPI) is a major partner of the green industy and is also dedicated to environmental concerns. They recognize both the global need to use water efficiently *and* the benefits of public and private green spaces. Through research, education and proper management, they believe that based on the following landscape water management principles 21st century landscapes can be increasingly water-efficient and meet the needs of the public and the environment alike.

■ Turfgrass is one of many important components of the landscape, providing numerous benefits and values to our quality of life, our environment and our eco-system.

■ The green industry in general and the turf industry in particular, play significant environmental and economic roles on the local and global levels.

■ There is no universal answer to all landscape water-management issues. Solutions need to be based on site-specific determinants, incorporating both initial establishment and continuous, long-term considerations.

■ Efforts to develop and implement any narrowly focused water-conservation solutions can prove problematic.

■ Efficient use of water use can be realized only through implementation of the combined best-management practices of the soil, plant, irrigation, landscape-maintenance and landscape-design sciences.

■ Actual water requirements of all landscape materials must be determined by means of objective and verifiable scientific processes, which in turn enable educated and environmentally sound landscape decisions.

■ Technological synergies, evolving from green-industry professionals and scientists will continue to expand and improve water-resource development, delivery, use and efficiency.

■ The public will take actions that simultaneously conserve water and improve the environment when properly informed of and motivated by the best available scientific knowledge and technology.

■ The basic right of individual artistic expression in the landscape and the value of a given plant is not solely determined by its need for and/or consumption of water.

■ Public policy should encourage the quality of life, the freedom of choice, and the emotional and economic values resulting from individually owned and public landscapes.